I0488408

Relation of Hydrologic Processes to Groundwater and Surface-Water Levels and Flow Directions in a Dune-Beach Complex at Indiana Dunes National Lakeshore and Beverly Shores, Indiana

By Paul M. Buszka, David A. Cohen, David C. Lampe, and Noel B. Pavlovic

In cooperation with the National Park Service

Scientific Investigations Report 2011–5073

U.S. Department of the Interior
U.S. Geological Survey

U.S. Department of the Interior
KEN SALAZAR, Secretary

U.S. Geological Survey
Marcia K. McNutt, Director

U.S. Geological Survey, Reston, Virginia: 2011

For more information on the USGS—the Federal source for science about the Earth, its natural and living resources, natural hazards, and the environment, visit http://www.usgs.gov or call 1–888–ASK–USGS.

For an overview of USGS information products, including maps, imagery, and publications, visit http://www.usgs.gov/pubprod

To order this and other USGS information products, visit http://store.usgs.gov

Suggested citation:
Buszka, P.M., Cohen, D.A., Lampe, D.C., and Pavlovic, N.B., 2011, Relation of hydrologic processes to groundwater and surface-water levels and flow directions in a dune-beach complex at Indiana Dunes National Lakeshore and Beverly Shores, Indiana: U.S. Geological Survey Scientific Investigations Report 2011-5073, 75 p.

Contents

Figures

Tables

Conversion Factors and Abbreviations

Multiply	By	To obtain
Length		
inch (in)	2.54	centimeter (cm)
inch (in)	25.4	millimeter (mm)
foot (ft)	0.3048	meter (m)
mile (mi)	1.609	kilometer (km)
Area		
square inch (in^2)	645.16	square millimeter (mm^2)
square mile (mi^2)	2.590	square kilometer (km^2)
Volume		
cubic foot (ft^3)	0.02832	cubic meter (m^3)
Flow rate		
foot per second (ft/s)	0.3048	meter per second (m/s)
foot per day (ft/d)	0.3048	meter per day (m/d)
cubic foot per day (ft^3/d)	0.02832	cubic meter per day (m^3/d)
inch per year (in/yr)	25.4	millimeter per year (mm/yr)
Hydraulic conductivity		
foot per day (ft/d)	0.3048	meter per day (m/d)

Temperature in degrees Fahrenheit (°F) may be converted to degrees Celsius (°C) as follows:

°C=(°F-32)/1.8

Vertical coordinate information is referenced to the National Geodetic Vertical Datum of 1929 (NGVD 29).

Horizontal coordinate information is referenced to the North American Datum of 1983 (NAD 83).

Altitude, as used in this report, refers to distance above the vertical datum.

Time is referenced to Eastern Standard Time.

Acronyms

INDU	Indiana Dunes National Lakeshore
NAD 83	North American Datum of 1983
NGVD 29	National Geodetic Vertical Datum of 1929
NPS	National Park Service
USGS	U.S. Geological Survey

Acknowledgments

The authors and the U.S. Geological Survey gratefully acknowledge the contributions of National Park Service (NPS) personnel from the Indiana Dunes National Lakeshore to this study, including those of Superintendent Constantine Dillon, Brenda Waters, Daniel Mason, Charles Morris, Robert Daum, and Laura Thompson, and NPS firefighting staff that assisted with well installation and surveying. We also gratefully acknowledge assistance to the project provided by representatives of the Town of Beverly Shores, the Shirley Heinze Land Trust, and many Beverly Shores residents who helped to obtain permissions to install wells or assisted with data collection.

The following USGS personnel also provided assistance to this project: Patrick Mills and Mark Hopkins (well installation), Lee Watson (retired; surface-water site installation), Kenneth Cole (1985 well installation at Howe's Prairie), and David Pollock (programming of the groundwater-flow analytical solution).

This report is Contribution 1646 of the USGS Great Lakes Science Center.

Relation of Hydrologic Processes to Groundwater and Surface-Water and Flow Directions in a Dune-Beach Complex at Indiana Dunes National Seashore and Beverly Shores, Indiana

By Paul M. Buszka[1], David A. Cohen[1], David C. Lampe[1], and Noel B. Pavlovic[2]

Abstract

The potential for high groundwater levels to cause wet basements (groundwater flooding) is of concern to residents of communities in northwestern Indiana. Changes in recharge from precipitation increases during 2006–9, water-level changes from restoration of nearby wetlands in the Great Marsh in 1998–2002, and changes in recharge due to the end of groundwater withdrawals for water supply since 2005 in a community at Beverly Shores, Ind., were suspected as factors in increased groundwater levels in an unconfined surficial aquifer beneath nearby parts of a dune-beach complex. Results of this study indicate that increased recharge from precipitation and snowmelt was the principal cause of raised water levels in the dune-beach complex from 2006 to 2009.

Annual precipitation totals in 2006–9 ranged from 43.88 to 55.75 inches per year (in/yr) and were substantially greater than the median 1952–2009 precipitation of 36.35 in/yr. Recharge to groundwater from precipitation in 2006–9 ranged from 13.5 to 22 in/yr; it was higher than the typical 11 in/yr because of large precipitation events and precipitation amounts received during non-growing-season months. An estimated increase in net recharge from reduced groundwater use in Beverly Shores since 2005 ranged from 1.6 in/yr in 2006 to 1.9 in/yr in 2009.

Surface-water levels in the wetland were as much as about 1.1 feet higher in 2007–9 (after the 1998–2002 wetland restoration) than during seasonally wet periods in 1979–89. Similar surface-water levels and ponded water were likely during winter and spring wet periods before and after wetland restoration. High water levels similar to those in 2009 were measured elsewhere in the dune-beach complex near a natural wetland during the spring months in 1991 and 1993 after receipt of near record precipitation. Recharge from similarly

high precipitation amounts in 2008–9 was also a likely cause of high groundwater levels in other parts of the dune-beach complex, such as at Beverly Shores.

Perennial mounding of the water table in the surficial aquifer indicates that the recharge that created the water-table mound originates within the dune-beach complex and not through flow from the adjacent hydrologic boundaries: the restored wetland, Lake Michigan, and Derby Ditch. Infiltrating precipitation causes most seasonal and episodic rises in groundwater levels beneath the dune-beach complex. Groundwater-level fluctuations lasting days to weeks in the dune-beach complex in 2008–9 were superimposed on a seasonal high water-table altitude that began with the recharge from snowmelt and rain in February 2009 and maintained through July 2009. Increases in water-table-mound altitude under the dune-beach complex recurred in 2008–9 in response to the largest rain events of 1 inch or more and to snowmelt. Smaller, shorter-term rises in water level after individual rain events persisted over hours to less than 1 week. Groundwater-level fluctuations varied over a relatively narrow range of about 2 to 3 feet, with no net fluctuations greater than 4 feet. Groundwater levels in or near low parts of the dune-beach complex were frequently within 0 to 6 feet of the land surface and indicate the potential for groundwater flooding.

Groundwater-level gradients from the water-table mound to wells next to surface-water discharges increase after rainfall and snowmelt events and recede slowly as groundwater discharges from the aquifer. Evapotranspiration is responsible for part of the general pattern of decreasing water-table altitudes observed from May to August 2009.

Rapid water-level rises in the restored wetland after precipitation do not likely have an effect on groundwater flooding elsewhere in the dune-beach complex. Surface-water-level fluctuations during this study generally varied over a narrower range, approximately from 1 to 1.5 feet, as compared with groundwater fluctuations, except after a very large, 10.77-inch rainfall. Time-delayed and smaller groundwater-level rises in wells near the restored wetland indicate a hydraulic delaying effect from the lower hydraulic conductivity of organic deposits in the subsurface near the marsh.

[1] U.S. Geological Survey Indiana Water Science Center, Indianapolis, Ind.

[2] U.S. Geological Survey Great Lakes Science Center, Lake Michigan Ecological Research Station, Porter, Ind.

Results of a simplified, steady-state cross-sectional model of groundwater flow also indicate that increased recharge from precipitation and snowmelt was the principal cause of raised water levels in the dune-beach complex from 2006 to 2009. Rises in the simulated water-table crest caused by increased recharge from precipitation in 2006–9 ranged from about 2 to 4 feet. A simulated addition of 1.9 in/yr of recharge from the water supply change raised the crest of the water-table mound by about 0.7 foot at about 900 feet from the restored wetland. The simulated groundwater-level change from a wetland water-level increase was generally smaller than that caused by precipitation and water-supply-derived recharge. The effect of a 1.1 foot simulated increase in water level in the restored marsh diminished to about a 0.75 foot groundwater level increase at about 900 feet from the marsh and to about a 0.55 foot groundwater level increase at about 1,500 feet from the marsh. Actual groundwater-level changes from wetland water-level increases would be smaller than simulated values because the organic sediments separating the wetland and the surficial aquifer tend to delay the response of groundwater levels to recharge and surface-water-level changes.

Introduction

From 2007 through 2010, the U.S. Geological Survey (USGS), in cooperation with the National Park Service (NPS), investigated the effect of natural and human-affected hydrologic processes on changes in groundwater and surface-water levels and groundwater-flow directions in nearby parts of a dune-beach complex at the Indiana Dunes National Lakeshore (INDU) and Beverly Shores, Ind., near Lake Michigan (fig. 1). Area residents are concerned about the altitude of the water table in the dune-beach complex relative to residential basement floors, as well as the processes affecting development of high groundwater levels and wet basements (groundwater flooding). The USGS and NPS wished to understand whether changes in precipitation, changes in water supply, and water-level changes in restored wetland areas such as the Great Marsh affect groundwater levels in a surficial aquifer under an adjacent dune beach-complex and its residential areas.

Ditch-drained interdunal wetlands adjacent to residential communities are common in northwestern Indiana and along the shores of the Great Lakes in the Midwestern States. The Great Marsh is a large interdunal wetland south of the dune-beach complex that parallels the modern Lake Michigan shoreline in northwestern Indiana (Shedlock and others, 1994, p. 6–7, fig. 2). During the 20th century, farming, ditching, and the construction of roads, levees, houses and other facilities reduced the size of Great Marsh from 12 mi to only 10 mi in length; the marsh width averages about 0.5 mi (National Park Service, 2009). Since 1998, the NPS has restored parts of the Great Marsh inside INDU to increase native-plant, migratory-bird, and other biological diversity and habitat (thereby protecting rare species), to increase recreational opportunities,

and to intercept runoff and slow its release to Lake Michigan and beaches along the lake (National Park Service, 2009).

Two conceptual diagrams illustrate typical interactions of precipitation and recharge with changes in groundwater levels in a hypothetical unconfined sand aquifer (figs. 2 and 3) similar to the surficial aquifer under the dune-beach complex between the Great Marsh and Lake Michigan (Shedlock and others, 1994). Precipitation that falls on the land surface, minus losses from evaporation and plant transpiration (evapotranspiration), can pond at the land surface, run off into surface water, or infiltrate through the unsaturated zone to groundwater as recharge (fig. 2). Another process that can contribute to recharge is the discharge of water from septic systems or from irrigation around homes and common areas. A building and its basement can be vulnerable to infiltration and groundwater flooding of the basement if the water table rises above the basement floor or sump underdrain and the sump pump is unable to withdraw sufficient water to lower the groundwater level below basement floors and maintain dry conditions (fig. 2).

Outflow from the surficial aquifer shown in figure 3 occurs in the form of groundwater seepage directly to ditches, wetlands, and to Lake Michigan (fig. 3, diagram *A*) and indirectly to those surface-water bodies through ditch or subsurface drains (fig. 3, diagram *B*), or to the atmosphere by seasonal processes such as transpiration (fig. 2). Withdrawals from the aquifer occur when water is pumped from shallow wells in the surficial aquifer for domestic supply and lawn irrigation or by sump pumps used for dewatering basements and areas near building foundations (fig. 2, diagram *B*). Sump withdrawals of groundwater are typically discharged within the same property, so these withdrawals may be essentially recycled. Withdrawals from sumps would lower groundwater levels near the building and slightly raise groundwater levels where the water is discharged but would produce little overall change in groundwater levels or flow (fig. 2).

A water-table altitude can rise because of increased amounts of recharge or decreased amounts of groundwater outflow. Recharge can increase when more precipitation falls or when water brought into an area increases and infiltrates to the water table. Discharge of groundwater from the hypothetical aquifer can decrease if flow to a tile drain is reduced by clogging or collapse or if the water-table gradient decreases in response to short-term (days or weeks) or long-term (months to years) water-level changes at a groundwater discharge such as a lake, ditch, or wetland (fig. 3, diagram *C*). The rate and volume of groundwater flow through a porous medium, such as in an unconfined sand aquifer, are directly proportional to the slope of the water-level surface (gradient) and the conductive characteristics of the aquifer to water (hydraulic conductivity).

Recent increases in annual precipitation in northwestern Indiana may have also increased surface-water levels in low-lying natural or restored wetlands and recharge to groundwater and groundwater levels under upland dune-beach complexes and associated residential areas. For example, the annual

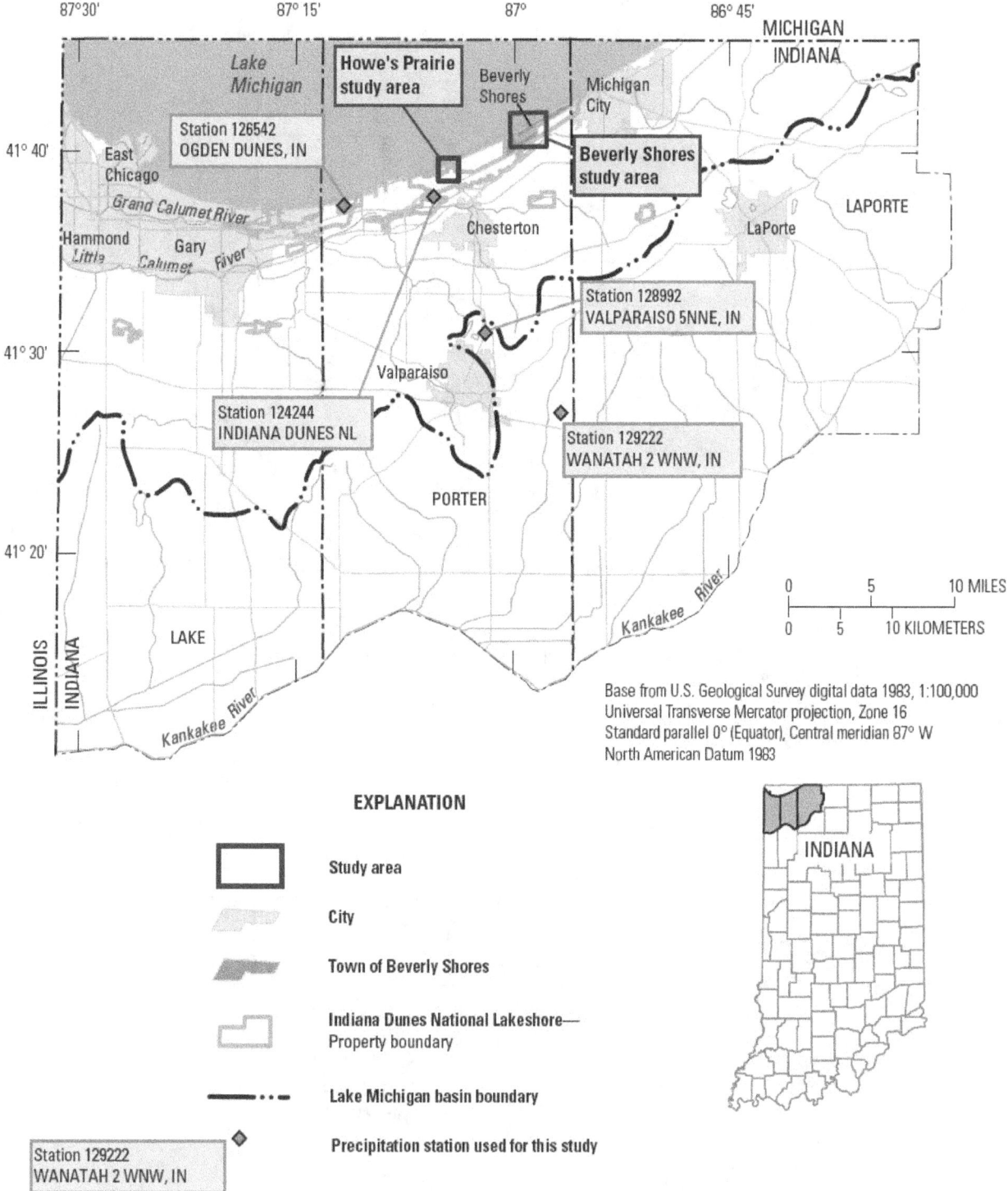

Figure 1. Location of Beverly Shores, Howe's Prairie, and nearby weather stations near the Indiana Dunes National Lakeshore, northwestern Indiana.

A. Conceptual diagram of parts of the water budget under the dune-beach complex, pre-2005 conditions

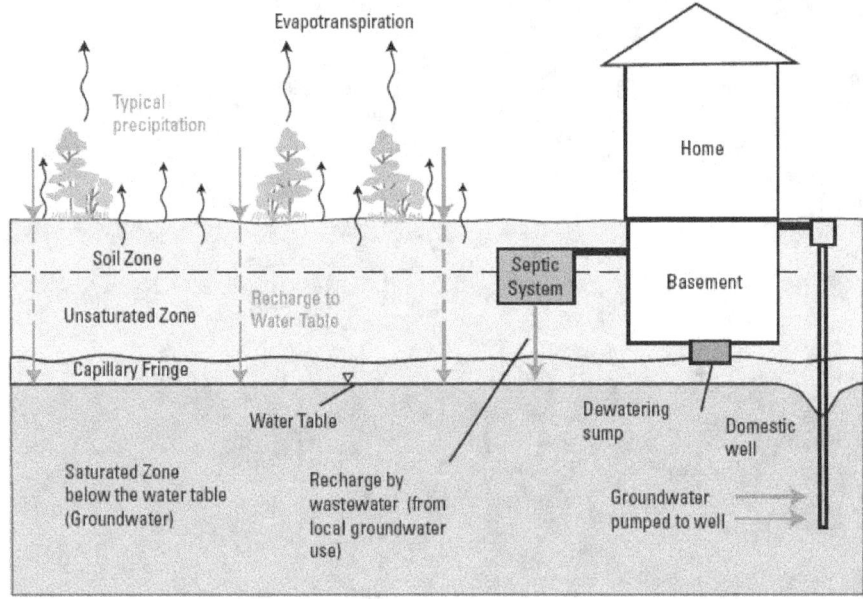

B. Conceptual diagram of parts of the water budget under the dune-beach complex, during 2005-9 conditions

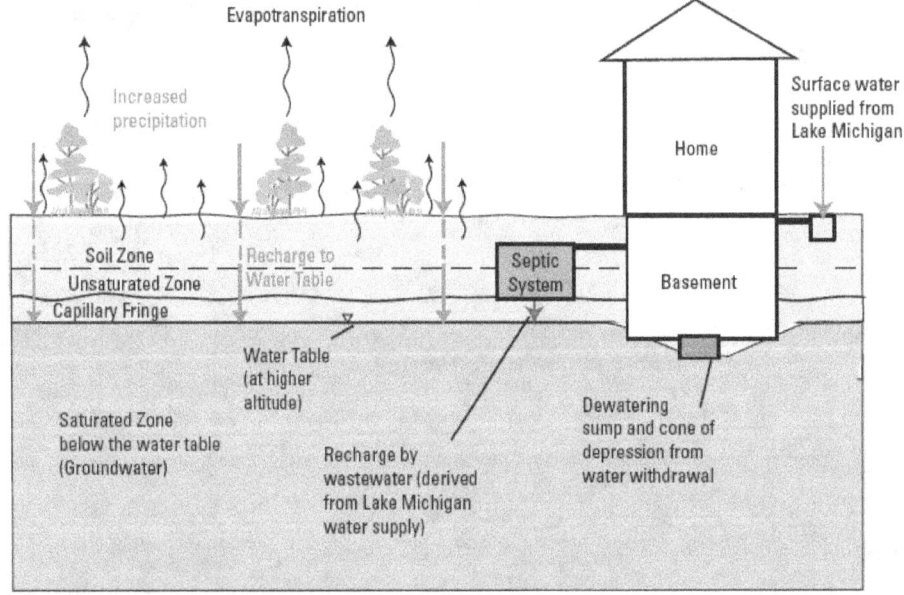

Diagrams modified from Alley and others (1999, fig. 4, p. 6).

Figure 2.   Interactions of precipitation and water-supply changes with recharge and groundwater levels in a hypothetical unconfined sand aquifer. (Note that water supply in *A* is local groundwater (pre-2005 conditions) and in *B* is water from Lake Michigan (2005–9 conditions).

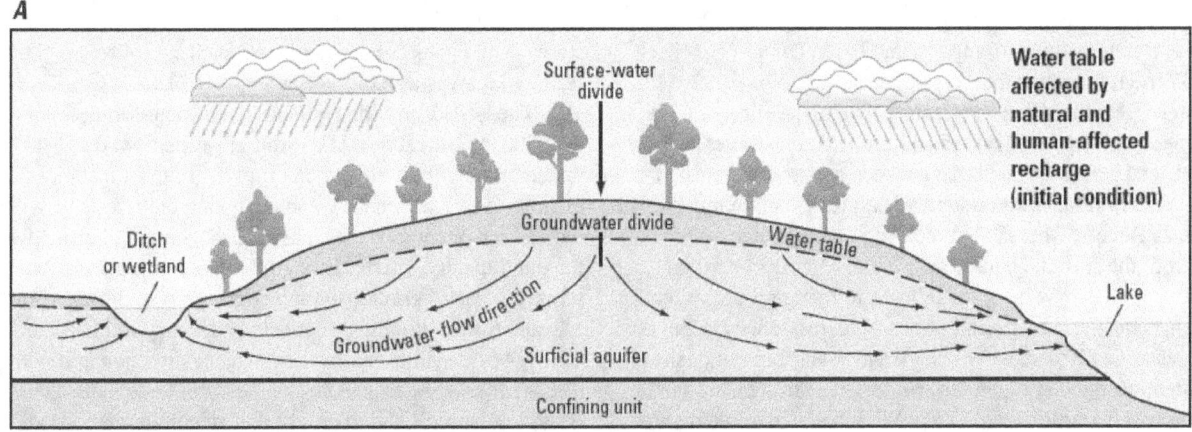

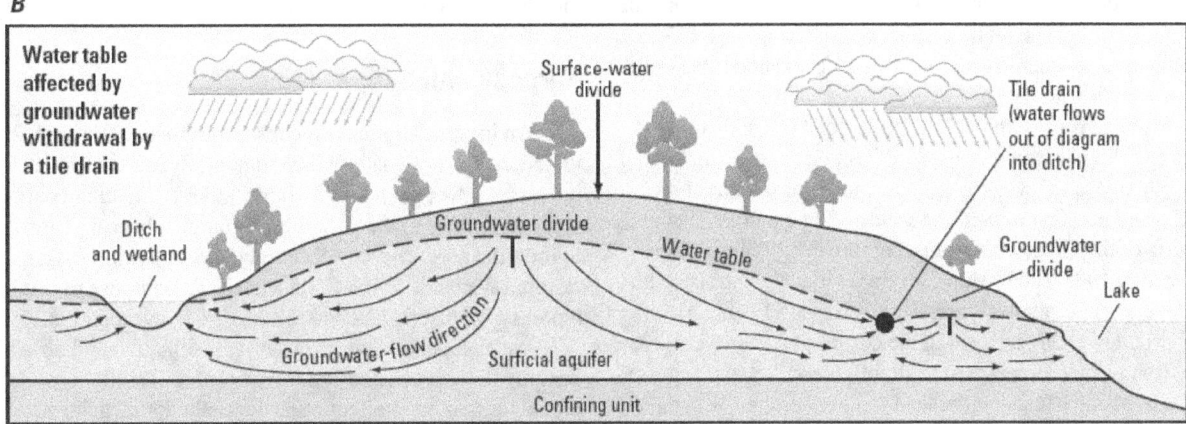

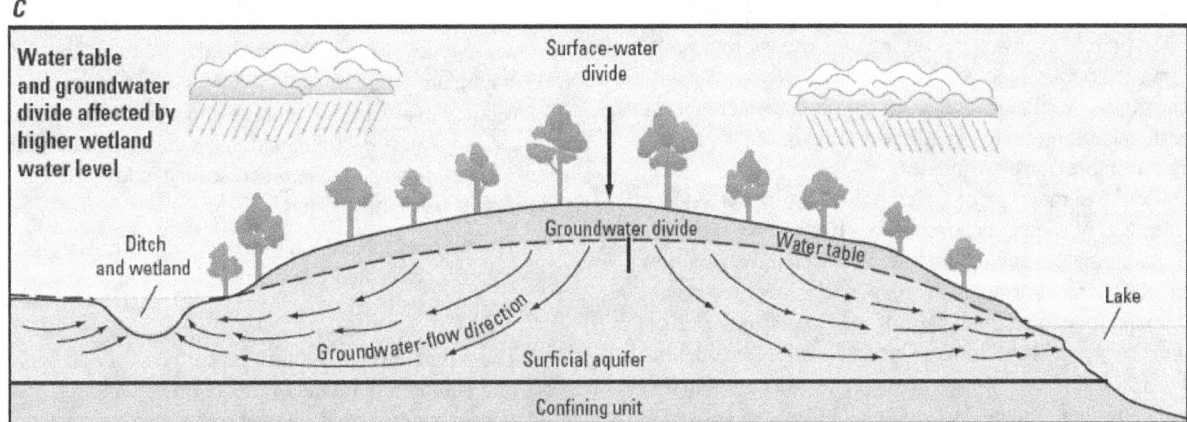

Diagrams reproduced and modified from
Grannemann and others (2000, fig. 5, p. 5).

Figure 3. Generalized groundwater flow: *A*, Under natural conditions; *B*, affected by tile drain flow; and *C*, affected by surface-water-level change in adjacent discharge (ditch-wetland). (Note that surface-water and groundwater divides coincide in *A* but not in *B* or *C*).

average precipitation total (rain and snow) at the INDU station (Indiana Dunes NL, fig. 1) in 2006 was 50.11 in/yr, substantially more than the average annual precipitation at that station (37.86 in/yr., the mean total precipitation for 1990–2000) and at a station about 10 mi west at Ogden Dunes (fig. 1; 36.37 in/yr, the mean total precipitation for 1970–1989; Midwestern Regional Climatic Center, 2010). The effect of these changes in precipitation on recharge and on groundwater levels at INDU and in adjacent areas has not been previously evaluated.

Recent (2005) water-supply changes in communities, such as Beverly Shores, that replaced groundwater withdrawals from the surficial aquifer with Lake Michigan-derived water supplies may have contributed to rising groundwater levels beneath those areas. For example, residences in Beverly Shores typically dispose of their wastewater through onsite residential septic systems. The net effect of decreasing well withdrawals and importing Lake Michigan water into an area where it ultimately recharges groundwater through septic-system discharge could increase water levels in the underlying surficial aquifer. The relative effect of the change to a Lake Michigan water supply on groundwater levels under the dune-beach complex has not been previously evaluated.

Wetland restoration at INDU included plugging culverts and ditches formerly used to drain wetlands, constructing spillways and levees to maintain surface water at higher levels, and planting of wetland species. For example, wetlands in parts of the Great Marsh adjacent to Beverly Shores were restored in 1998-2002 behind several spillway controls south of Beverly Drive and east and west of Broadway (fig. 4). The structure south and east of the corner of Beverly Drive and Broadway was designed to slightly raise wetland water levels to about 1 ft above the land surface (Brenda Waters, National Park Service, written commun., 2010). The former course of Derby Ditch in these areas was plugged where it crossed old roadbeds and spillways were installed to minimize retention time of excess water during and after storm events and limit temporary blockages from natural events. The net effect of raised wetland water levels on groundwater levels in the surficial aquifer under an adjacent dune-beach complex has not been previously evaluated.

Additional mapping of water-table and surface-water altitudes was needed to determine the effects of recent precipitation, water-supply changes, and wetland restoration on the directions of groundwater flow under the dune-beach complex relative to restored parts of the Great Marsh. Large, perennial water-table mounds of broad extent—several hundred to 1,500 ft wide—were mapped in October 1980 in the surficial aquifer in dune-beach complex sediments between the Great Marsh and Lake Michigan (Shedlock and others, 1994, p. 34–35). Groundwater in the surficial aquifer flowed generally from recharge areas in the dune-beach complex to the adjacent ditches, wetlands, and Lake Michigan (Shedlock and others, 1994). Smaller, transient water-table mounds lasting from days to weeks after precipitation were mapped under a topographic depression near the margin of the dune-beach complex about 0.8 mi west of the study area. The localized,

transient water-table mounds under the depression temporarily reversed groundwater flow directions (Shedlock and others, 1993; Shedlock and others, 1994, figs. 14 and 15, p. 36–37). Few data were available from those studies to understand water-level changes beneath dune-beach complexes in the residential areas at Beverly Shores.

The USGS provides reliable scientific information to describe the interaction of hydrologic systems and assists in understanding their effects on property. In most instances, this involves documenting and analyzing the effects of widely recognized phenomena such as surface-water and groundwater flow and quality. On occasion, this involves relatively small-scope studies of previously unrecognized phenomenon, such as groundwater flooding, that emerge as a natural hazard. The USGS works within its strategic science direction and with its cooperative partners, such as the NPS, to document these emerging hazards and to ensure that scientific methods are applied effectively to better understand these phenomena and thereby minimize loss.

Purpose and Scope

An investigation was completed from 2007 through 2010 to understand how natural and human-affected hydrologic processes affect shallow groundwater levels in an unconfined surficial aquifer in the dune-beach complex beneath Beverly Shores and surface-water levels and directions of flow in restored wetlands in the Great Marsh. The purpose of this report is to present data and results of the investigation.

Water-level and weather data (precipitation and air temperatures) are presented and interpreted to identify hydrologic processes that most affect water levels in the surficial aquifer. Long-term variations in groundwater levels were also evaluated at an undeveloped, natural site in the dune-beach complex at Howe's Prairie. The intent was to understand whether groundwater-level changes in dune-beach-complex areas at INDU may relate predominantly to changes in precipitation.

The study evaluates the relative importance of factors that contributed to water-level fluctuations in the surficial aquifer during 2008–9; namely, those processes that add water to or remove water from the aquifer:

- The ways in which precipitation and water-supply changes translated into changes in recharge to the surficial aquifer,

- The possible relation of raised surface-water levels and perennially ponded water from wetland restoration in the Great Marsh to changes in groundwater and surface-water levels in adjoining residential areas of Beverly Shores, and

- Changes in water supply, from tapping of local groundwater to importation of water into the community from a Lake Michigan-based supply.

The study also compares groundwater levels with base altitudes of hypothetical basements to identify the potential

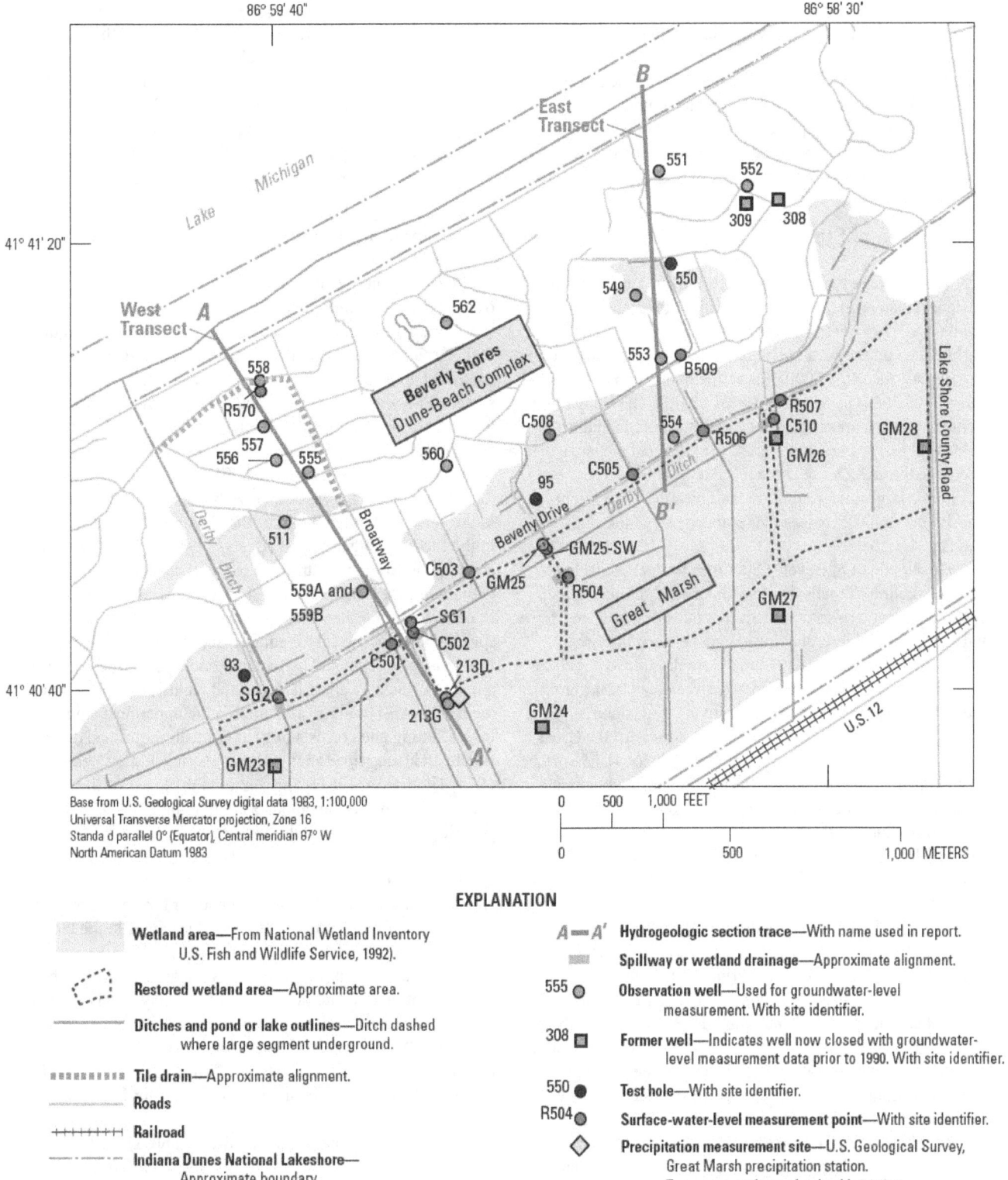

Base from U.S. Geological Survey digital data 1983, 1:100,000
Universal Transverse Mercator projection, Zone 16
Standard parallel 0° (Equator), Central meridian 87° W
North American Datum 1983

EXPLANATION

Wetland area—From National Wetland Inventory
U.S. Fish and Wildlife Service, 1992).

Restored wetland area—Approximate area.

Ditches and pond or lake outlines—Ditch dashed
where large segment underground.

Tile drain—Approximate alignment.

Roads

Railroad

Indiana Dunes National Lakeshore—
Approximate boundary.

A — A'   Hydrogeologic section trace—With name used in report.

Spillway or wetland drainage—Approximate alignment.

555   Observation well—Used for groundwater-level
measurement. With site identifier.

308   Former well—Indicates well now closed with groundwater-
level measurement data prior to 1990. With site identifier.

550   Test hole—With site identifier.

R504   Surface-water-level measurement point—With site identifier.

◇   Precipitation measurement site—U.S. Geological Survey,
Great Marsh precipitation station.
Temporary station maintained in 2008-9.

Figure 4.   Part of study area at Beverly Shores, Indiana, including hydrogeologic sections *A–A'* and *B–B'*, wells, and surface-water
sites with data used for this investigation near the Great Marsh at Indiana Dunes National Lakeshore, northwestern Indiana.

for seasonal (months) and episodic (days to weeks) ground-water flooding of basements. The most important processes that affected groundwater-level changes during the study were identified by using an analytical model to simulate the interactions of changes in precipitation, water use, and adjacent wetland water levels in a simplified model of the surficial aquifer under the dune-beach complex.

Description of Study Area

The study area is in northeastern Porter County, northwestern Indiana (fig. 1). The investigation focused principally on about a 1.7-mi² area in Beverly Shores and INDU between Derby Ditch on the west and about County Line Road on the east (fig. 4). As of the 2000 census, the population of Beverly Shores was 708 (U.S. Census Bureau, 2010). The study area included parts of areas drained by Derby Ditch, including residential parts of the dune-beach complex and wetlands that discharge through Derby Ditch to Lake Michigan. The Great Marsh, the largest interdunal wetland near the Lake Michigan shore at INDU, crosses the southern part of the study area (fig. 4).

Development of Beverly Shores in the late 1920s led to the construction of roads, low embankments, drainage pipe, ditches, and culverts to drain adjacent parts of the Great Marsh and the dunes and interdunal wetlands between the Great Marsh and Lake Michigan. Homes were developed largely in the dunes and interdunal wetlands between the Great Marsh and Lake Michigan (referred to locally as "the Island"), although some were built in or on the periphery of the Great Marsh. Many of the homes were built with below-land-surface basements that are equipped with sump pumps to maintain a dry subsurface space. When Congress authorized INDU in the mid-1960s, much of the acreage in Beverly Shores between highway U.S. 12 and the residential part of the dune-beach complex was included within INDU. Areas along U.S. 12 and in the Island were later excluded from INDU.

Land use in the study area is principally residential along streets in Beverly Shores and is a mixture of residential and commercial uses along U.S. 12. Principal transportation land uses that cross the study area consist of U.S. 12, paved and unpaved secondary roads, and railroads. Parts of the study area in INDU and adjacent parkland are maintained as natural and restored wetlands, wooded dune, and swale environments. Howe's Prairie is an area of about 0.1-mi² within the dune-beach complex of INDU and about 5 mi west-southwest of Beverly Shores (fig. 5). Howe's Prairie is a panne: a bowl-shaped, groundwater fed, flat-bottomed basin that originated soon after the Tolleston-age dunes were formed about 4,000 years ago when blowouts (depressions) in the dune were excavated by the wind down to the water table (Thompson, 1992; Cole and Taylor, 1995). Variable water levels and land-surface slopes within Howe's Prairie host a variety of wetland and other prairie settings: fen, swamp, wet, moderately wet and dry sand prairie, oak savanna, and oak forest vegetation (Cole

and Taylor, 1995). Part of a water-table mound in the surficial aquifer is directly beneath Howe's Prairie in the dune and swale topography between the Great Marsh and Lake Michigan (Shedlock and others, 1994). The Great Marsh and Dunes Creek are south of Howe's Prairie and cross the central part of that part of the study area from northeast to southwest. Howe's Prairie is undeveloped, and no artificial drainage is present.

Hydrogeologic Framework

Groundwater resources in the study area chiefly consist of two regionally important, unconsolidated aquifer systems (the Calumet Aquifer System and the Lacustrine Plain Aquifer System; fig. 6) and one underlying bedrock aquifer system (the Silurian-Devonian Carbonate Aquifer System). (All aquifer-system names are local usage; see Beaty, 1994.) The Calumet and Lacustrine Plain Aquifer Systems were the most extensively used for groundwater withdrawals by residences and for public supply in the area before 2005.

The Beverly Shores and Howe's Prairie study areas are in the surficial aquifer, a part of the Calumet Aquifer System in the Lake Michigan region (Beaty, 1994), and in dune-beach complex and wetland settings (fig. 6 and table 1). The hydrogeologic framework of the Great Marsh area of Porter County in northwestern Indiana consists of three glacial aquifers—the surficial aquifer, the subtill aquifer, and the basal sand aquifer—typically separated by variable thicknesses of glacial till and lacustrine clay and silt (fig. 7 and table 1; Shedlock and others, 1994). The surficial aquifer is the uppermost aquifer in this setting and consists of fine- to medium-grained dune, beach, and lacustrine sands and gravels (fig. 7, also Beaty, 1994; Thompson, 1987). The saturated thickness of the surficial aquifer ranges from about 5 ft along the Lake Michigan shoreline to 30 to 35 ft in the areas south of Beverly Shores near highway U.S. 12 (Watson and others, 2002; Shedlock and others, 1994, fig. 9, p. 25) and near areas drained by Derby Ditch. Groundwater in the surficial aquifer typically flows from the dune-beach complex toward discharges in the Great Marsh, ditches, and Lake Michigan (fig. 7).

The surficial aquifer generally is unconfined but can be locally confined by interlaminated silt and clay, marls, calcareous mud, and peat (organic) deposits (Shedlock and others, 1994, p. 17). Confining units of interbedded till and glacial lacustrine clay and silt separate the surficial aquifer from the underlying subtill aquifer and the subtill aquifer from the basal sand aquifer.

The subtill and basal sand aquifers (table 1) are assumed to be equivalent to the Lacustrine Plain Aquifer System of Beaty (1994; fig. 6). The subtill aquifer in the study area is a series of sand units with interbedded lenses of clay that underlies the confining unit beneath the surficial aquifer near the Lake Border Moraine (fig. 7; table 1), principally south of the study area. The saturated thickness of the subtill aquifer in the study area ranges from nonexistent to about 80 ft south of the study area (Shedlock and others, 1994, fig. 22, p. 52).

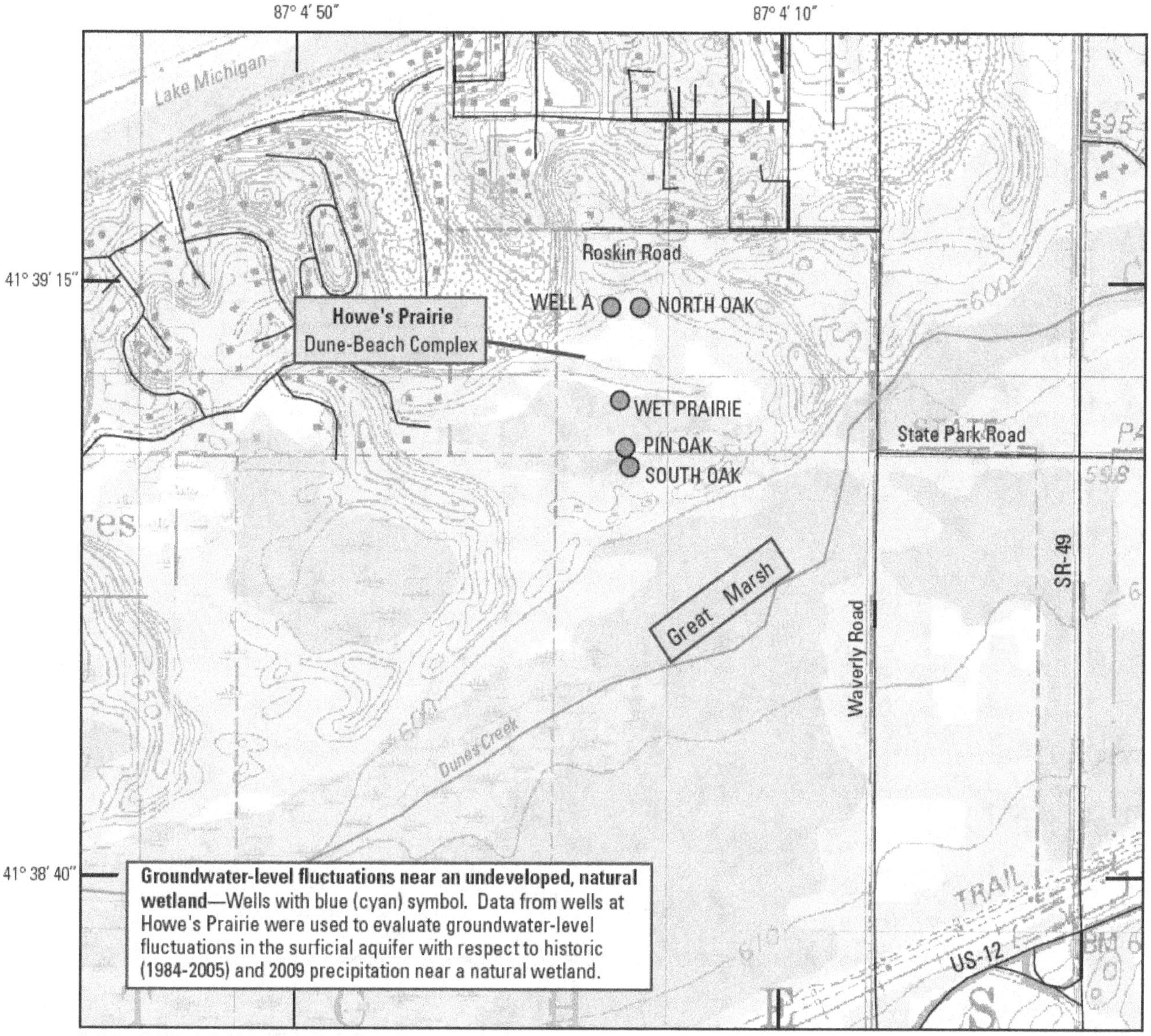

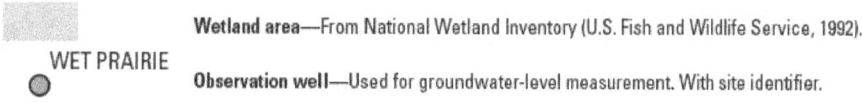

Figure 5.   Part of study area at Howe's Prairie, including wells with data used for this investigation near the Great Marsh at Indiana Dunes National Lakeshore, northwestern Indiana.

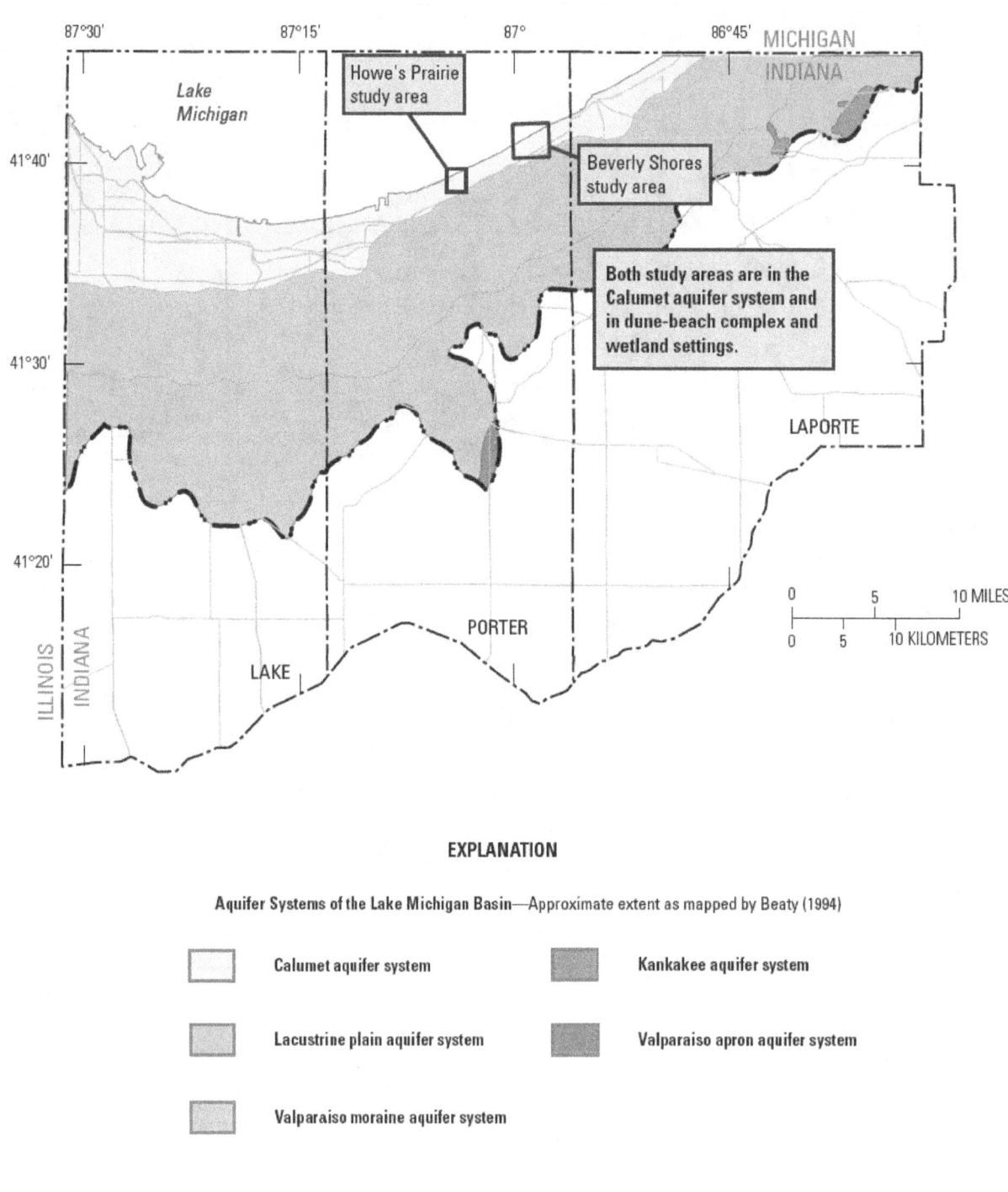

EXPLANATION

Aquifer Systems of the Lake Michigan Basin—Approximate extent as mapped by Beaty (1994)

Calumet aquifer system

Kankakee aquifer system

Lacustrine plain aquifer system

Valparaiso apron aquifer system

Valparaiso moraine aquifer system

— ·· — Lake Michigan basin boundary

Figure 6. Unconsolidated aquifer systems in the Lake Michigan region and the study areas, northwestern Indiana. Definitions of aquifer-system characteristics can be found in Beaty (1994, pl. 2.)

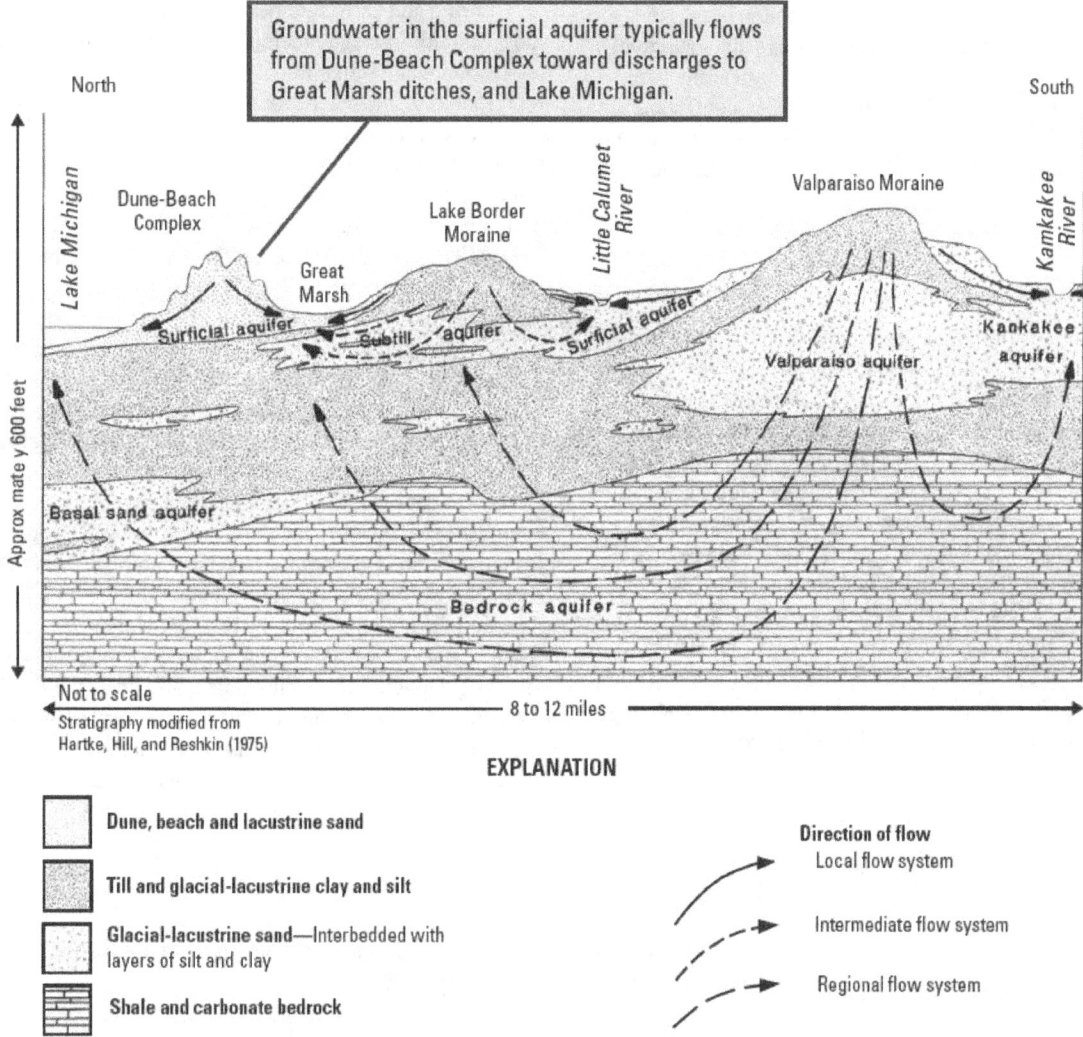

Figure 7. Diagrammatic hydrogeologic section showing aquifers and conceptual groundwater-flow directions in the eastern half of the Indiana Dunes National Lakeshore near Beverly Shores, northwestern Indiana. Diagram reproduced from Shedlock and others (1994, fig. 7, p. 21)

Table 1. Comparison of hydrogeologic framework of the study areas, northwestern Indiana, with those of previous investigations.

Lithostratigraphic descriptions-	Hydrogeologic-framework designation	
	This report and Shedlock and others (1994)	Beaty (1994)
Fine- to medium-grained dune, beach, and lacustrine sands and gravels (of Holocene and Pleistocene age)	Surficial aquifer	Calumet aquifer system
Glacial and lacustrine sands of Pleistocene age, with interbedded clays and silts, including tills of the Lake Border Moraine	Confining units	Lacustrine Plain aquifer system
	Subtill aquifer	
	Confining units	
	Basal sand aquifer	
Shale and carbonate rocks of Devonian and Silurian age	Bedrock aquifer	Silurian-Devonian carbonate aquifer system

A deeper, confined sand aquifer, known as the basal sand aquifer, consists of discontinuous lenses of sand and gravel interbedded with layers of silt and clay at or near the base of unconsolidated glacial deposits where they contact the bedrock (fig. 7). Wells that flow at land surface in the local area were completed in the basal sand aquifer (Shedlock and others, 1994, p. 21).

Climate

The climate in Indiana has four distinct seasons, frequently cold winters, a transitional spring season with active precipitation, hot humid summers followed by typically drier, cooler autumns (Scheeringa, 2002). Climate normals, the average of a climate element such as temperature over a 30-year interval, are available from two stations near the study area: Ogden Dunes for 1970 through 1989 data, and INDU headquarters for 1989 through 2000 data (Midwestern Regional Climate Center, 2010; table 2). The Ogden Dunes station was discontinued on May 21, 1989 and relocated to the present INDU site, where data collection resumed on June 1, 1989. Climate statistics based on combined data from the two stations from 1970 through 2000 are similar (table 2) and, therefore, the two records are used in this study as one continuous record.

Spring and early summer months typically have the most rainfall. Average precipitation can range from about 3 to less than about 5 in/month from April through November, as represented by the record for the weather station at Valparaiso (Midwestern Regional Climate Center, 2010; station location in fig. 1). Winter months are typically drier, except near Lake Michigan. Lake Michigan affects the study-area climate; air temperature extremes are moderated by the proximity to Lake Michigan (Scheeringa, 2002). The average date of the last freezing temperature in spring in extreme northwestern Indiana is about May 1 near Lake Michigan. The study area is also typically in the path of "lake-effect" snowfall. The heaviest snowfalls in Indiana are in the northern part of the State, including the study area, near and east of the southern tip of Lake Michigan (Scheeringa, 2002). Lake-effect snow is produced during the autumn and winter when cold, arctic air masses move across long fetches of the relatively warmer open water of Lake Michigan (Changnon, 1968). Moist air from the warm lake rises into the cold, arctic air where it then cools and condenses into snow clouds. The prevailing wind direction through the depth of the snow clouds determines where the snow will fall, and wind duration and intensity affect how much snow accumulates. Soil moisture conditions occasionally become saturated during the winter and spring with recharge from meltwater and precipitation and decline through the growing season (Scheeringa, 2002).

Table 2.   Minimum and maximum monthly average and annual average temperatures and precipitation for weather stations at Ogden Dunes, 1970–89 and Indiana Dunes National Lakeshore, 1989–2000.

[--, no data; F, Fahrenheit]

Property	Statistic	Months	Weather station	
			Ogden Dunes	**Indiana Dunes National Lakeshore**
Station identifier	--	--	126542 OGDEN DUNES, IN	124244 INDIANA DUNES NAT LKSHR, IN
Period of record for computing averages	--	--	1970 to 1989	1989 to 2000
Temperature, in °F	Minimum	January	24.4	22.6
	Maximum	July	73.6	72.5
	Annual Average	--	50.6	49.1
Precipitation, in inches	Minimum	February	1.62	1.58
	Maximum	June	4.41	4.64
	Annual Average	All	36.37	37.86
Snowfall, in inches	Minimum	Several	0	0
	Maximum	January	14.6	14.5
	Annual	All	44.7	41.4

Methods of Data Collection and Analysis

A network of wells and reference marks (where ground-water and surface-water levels were measured) and a rain gage (where precipitation was recorded) were established by this study. Methods used to establish vertical control on all measuring points are described in the section "Observation Wells and Surface-Water-Level Sites." Application of a cross-sectional analytical solution to analyze which hydrologic processes most affect changes in the water-table profile in a simplified version of the surficial aquifer is discussed later in this report.

Precipitation and Water-Use Data

Precipitation and air temperatures reported by this study were compiled from the two weather stations mentioned previously. Data were collected from October 1951 to April 1989 at the former weather station at Ogden Dunes, about 10 mi west of Beverly Shores (fig. 1; Station 126542, Midwestern Regional Climate Center, 2010). Data were collected from June 1989 through December 2009 at the weather station at INDU park headquarters, about 3.5 mi west of Beverly Shores (fig. 1; Station 124244, Midwestern Regional Climate Center, 2010; Laura Thompson, National Park Service, written commun., 2010). Records of rainfall were collected from November to December 2008 and March to December 2009 at a site

in the Great Marsh (Great Marsh station) about 500 ft east of well 213G by using a data-logging, tipping-bucket rain gage (fig. 4). Metered water use data from 2006–9 for the residential area at Beverly Shores were compiled and provided by the Michigan City Department of Water Works (Ron Plamowski, Michigan City Department of Water Works, written commun., 2009; Geof Benson, Town of Beverly Shores, written commun., 2009).

Observation Wells and Surface-Water-Level Sites

Two available observation wells and 13 new observation wells were used to measure groundwater levels for the part of the study area at Beverly Shores (fig. 4). The two available wells, identified as 213G and GM25 (fig. 4), were completed to depths of 12 and 8 ft, respectively (table 3). The 13 new wells were installed by using a USGS direct-push drill rig (shown in fig. 8) or a hand-operated well driver. Casing materials were a mix of black steel or galvanized steel pipe with threaded couplings. Well screens in 11 of the 13 new wells were constructed of 2-in.-inside-diameter wire-wrapped stainless-steel well points with 0.010-in. slots; the other two screens were perforated steel well points (0.125-in.-diameter perforations, 16 holes per square inch) that were backed with an 80 gauze galvanized wire mesh (80 holes per square inch). Well depths ranged from about 5 to 28 ft below land surface (table 3).

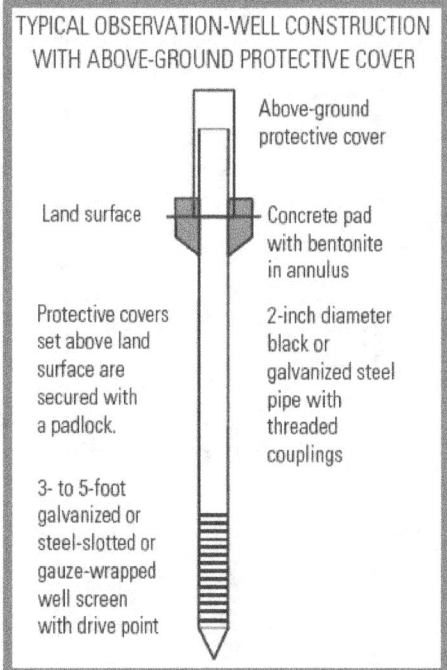

Figure 8. The direct-push drill rig used to collect sediment cores and construct observation wells for this study, a core collected with this drill rig of surficial aquifer sand from well 554, and a diagram of a typical observation well installation (photographs by David A. Cohen, U.S. Geological Survey, July 2008). Length of core in center image is about 3.5 ft.

Table 3. Characteristics of observation wells in the Great Marsh-Beverly Shores network, northwestern Indiana. Well locations are shown in fig. 4

[USGS, U.S. Geological Survey; ddmmss s, degrees, minutes, seconds; NGVD 29, National Geodetic Vertical Datum of 1929; SS, Stainless steel; WW, wire wrapped screen; Galv, Galvanized steel; Gauze, gauze screen overlay; Unless otherwise indicated, all casing and screen inside diameters are nominal 2-inch]

Well name on figure	USGS station-identification number	Latitude (North, in ddmmss.s)	Longitude (West, in dddmmss.s)	Date drilled (month/day/year)	Method of installation	"Land-surface datum (feet above NGVD 29)	Measuring-point altitude[1] (feet above NGVD 29)	Altitude accuracy (feet)	Casing material	Well-screen material	Well-screen length (feet)	Depth from land surface to bottom of well screen (feet below land-surface datum)
GM-25	414050086590501	414052.8	0865906.0	07/13/1979	Hand driven	596.7	599.54	0.01	SS	SS-WW	3	8
213G	414041086592003	414039.0	0865918.0	10/17/1979	Hollow-stem auger drill rig	600.5	600.5	.01	PVC	PVC-slotted	3	12
511	414056086593801	414056.0	0865938.4	05/15/2007	Hand driven	600.64	602.88	.02	Steel	SS-WW	3.1	5.1
549	414115086585401	414115.2	0865854.1	04/24/2008	Direct-push drill rig	604.62	606.16	.02	Steel	SS-WW	2.3	21.36
551	414126086585101	414126.3	0865851.1	07/28/2008	Direct-push drill rig	607.22	609.95	.02	Steel	SS-WW	2.2	14.77
552	414125086583901	414125.5	0865839.5	07/28/2008	Direct-push drill rig	620.04	622.8	.02	Steel	SS-WW	1.8	28.04
553	414109086585001	414109.5	0865850.8	07/28/2008	Direct-push drill rig	607.12	609.79	.02	Steel	SS-WW	2.2	15.03
554	414102086584801	414102.0	0865848.9	07/30/2008	Direct-push drill rig	600.8	603.3	.02	Steel	Galv-Gauze	2.5	16.5
555	414059086593401	414059.2	0865934.8	07/29/2005	Direct-push drill rig	609.04	612.18	.02	Steel	SS-WW	2.2	18.96
556	414101086593901	414101.2	0865939.5	07/31/2008	Direct-push drill rig	604.79	607.28	.02	Steel	SS-WW	2	20.71
557	414104086594101	414104.3	0865941.2	07/30/2008	Direct-push drill rig	609.57	612.28	.02	Steel	SS-WW	2.2	18.09
558	414108086594101	414108.3	0865941.7	07/31/2008	Direct-push drill rig	605.11	607.98	.02	Steel	SS-WW	2.2	14.93
559	414049086592801	414049.2	0865928.4	08/01/2008	Direct-push drill rig	599.62	602.11	.02	Steel	Galv-Gauze	2.5	11.61
560	414100086591801	414100.4	0865918.1	07/29/2008	Direct-push drill rig	607.44	610.37	.02	Steel	SS-WW	1.1	14.87
562	414112086591701	414112.6	0865917.9	07/29/2008	Direct-push drill rig	608.06	605.16	.02	Steel	SS-WW	2.2	11.6

Sediment cores were collected at five well sites and one test hole (fig. 4, wells 549, 554, 555, 557, and 558, and test hole 550) by hydraulically pushing a 4-ft-long coring tube with a 1.5-in.-diameter acrylic liner. Sediment cores were described according to sediment type, texture, and color and were used to select the depth of observation wells.

Various types of measuring points were established to measure surface-water levels at 14 sites (fig. 9; table 4). Available stable culvert ends were selected to establish reference marks if appropriately located relative to restored wetlands or other wetland areas and ditches. Staff gages (SG1 and

SG2) were installed in two restored wetland pools that were upstream from the controls near the southeastern corner of Beverly Drive and Broadway and at the Derby Ditch crossing under Beverly Drive (fig. 9).

Altitudes of measuring points used to measure ground-water and surface-water levels were established by an optical level survey and are referenced to the National Geodetic Vertical Datum of 1929 (NGVD 29). The accuracy of most measuring points was surveyed to within 0.02 ft. All horizontal locations established by this work were referenced to the North American Datum of 1983 (NAD 83).

Figure 9. Photograph showing typical surface-water measurement sites: (left) a stable culvert end at site CUL510 and (right) a staff gage at site SG1. Photographs by David A. Cohen, U.S. Geological Survey, (left) May 2007 and (right) August 2008.

Table 4. Characteristics of surface-water data-collection sites in the Great Marsh-Beverly Shores network, northwestern Indiana. Site locations are shown in fig. 4.

[USGS, U.S. Geological Survey; ddmmss.s, latitude in degrees, minutes, seconds; dddmmss.s, longitude in degrees, minutes, seconds; NGVD 29, National Geodetic Vertical Datum of 1929]

Site name on figure	USGS station-identification number	Formal site name	Latitude (North, in ddmmss.s)	Longitude (West, in dddmmss.s)	Measuring-point altitude (feet above NGVD 29)	Altitude accuracy (feet)
C501	414044086592501	Culvert 501 SW Beveryly Dr-Broadway at Beverly Shores	414044.0	0865924.8	595.08	0.2
C502	414045086592201	Culvert 502 SE Beverly Dr-Broadway at Beverly Shores	414045.1	0865922.3	596.91	.02
C503	414045086591501	Culvert 503 NE Beverly Dr-Pearson at Beverly Shores	414050.4	0865915.4	597.68	.02
R504	414045086590301	Rebar 504 Constance Control at Beverly Shores, IN	414049.9	0865902.9	600.83	.20
C505	414059086585501	Culvert 505 SW Beverly-St Clair at Beverly Shores, IN	414059.2	0865854.5	599.45	.02
R506	414103086584501	Rebar 506 SE Beverly-McAlister at Beverly Shores, IN	414103.1	0865845.4	600.6	.20
R507	414106086583601	Rebar 507 SE Beverly-Wells at Beverly Shores, IN	414105.8	0865835.8	601.93	.20
C508	414103086590501	Culvert 508 Idler NR Shore Dr at Beverly Shores, IN	414102.8	0865905.0	600.47	.20
B509	414110086584901	Bridge MP 509 north of Idler at Beverly Shores, IN	414109.9	0865848.6	605.88	.02
C510	414104086583601	Culvert 510 SE Beverly-Wells at Beverly Shores, IN	414103.8	0865836.8	600.67	.20
R570	414108086594102	Rebar 570 South Fairwater at Beverly Shores, IN	414108.0	0865942.0	602.22	.02

Groundwater- and Surface-Water-Level Measurements

Weekly and more frequent occasional groundwater and surface-water levels were measured by National Park Service staff at most wells and measuring points, using USGS-defined protocols. Water levels in wells and at surface-water gages were monitored from September 2008 through November 2009 at least weekly and, on occasion, more frequently in response to substantial rainfall from May 2007 through November 2009 (fig. 4). Additional records of groundwater levels were collected during this period at 15-minute intervals using transducers equipped with data loggers in wells along section *B–B′* extending from adjacent to the Great Marsh (well 554) and north across the dune-beach complex (wells 553 and 551), and in one well on section *A–A′* (well 556). All water-level data were archived in the USGS National Water Information System.

Historical records of occasionally measured groundwater and surface-water levels were compiled for 10 wells and 2 measuring points in the Beverly Shores part of the study area by using files of the USGS Indiana Water Science Center. Data from two wells installed during past NPS/USGS projects (fig. 4, wells GM25 and 213G) were used to monitor groundwater levels near the restored wetlands before and after restoration. Although some other wells such as 308 and 309 were no longer present as of the beginning of this study (fig. 4), their water levels were compared with those from nearby sites with similar hydrologic settings and conditions. Comparisons involving specific wells and measuring points are discussed later in this report.

Groundwater levels from five wells in an interdunal wetland at Howe's Prairie also were used to compare groundwater-level responses at Beverly Shores to precipitation changes in a hydrogeologically similar area that was not affected by residential development or wetland restoration. Howe's Prairie is about 5 mi southwest of Beverly Shores and is affected by the same regional precipitation as Beverly Shores. The Howe's Prairie wells were installed in 1985 with a hand-operated well driver; they consisted of 2-in.-inside-diameter steel pipe with well-point-type well screens. Water levels were measured monthly from April 1985 through 2005 except during winter months, when ice in the well obstructed measurements, and then intermittently from April through September 2009. Information about the wells at Howe's Prairie is listed in table 5. The name convention for the local well identifiers and reference data established for the Howe's Prairie wells by Dr. Kenneth Cole (U.S. Geological Survey, Flagstaff, Arizona) and one of the authors (Dr. Noel Pavlovic) were generally maintained by this study.

Table 5.   Characteristics of observation wells at Howe's Prairie near the Great Marsh, northwestern Indiana. Well locations are shown in fig. 5.

[USGS, U.S. Geological Survey; ddmmss s, latitude in degrees, minutes, seconds; dddmmss.s, longitude in degrees, minutes, seconds; NGVD 29, National Geodetic Vertical Datum of 1929; Unless indicated otherwise, all casing and screen diameters are nominal 2-inch]

Local well name	USGS station-identification number	Latitude (North, in ddmmss.s)	Longitude (West, in dddmmss.s)	Date of first measurement (Month and year)	Method of installation	Land-surface datum (feet above NGVD29)	Measuring-point altitude[1] (feet above NGVD29)	Altitude accuracy (feet)	Casing material	Well-screen material	Well-screen length (feet)
WELL A	413904087042401	413904.1	0870424.3	04/1985	Hand driven	604.49	604.94	0.1	Steel	Drive point	2 to 3
NORTH OAK	413913087042201	413913.5	0870422.0	04/1985	Hand driven	606.91	608.66	.1	Steel	Drive point	2 to 3
WET PRAIRIE	413908087042401	413908.0	0870423.5	04/1985	Hand driven	601.54	601.74	.1	Steel	Drive point	2 to 3
PIN OAK	413905087042301	413905.3	0870423.1	04/1985	Hand driven	601.51	602.13	.1	Steel	Drive point	2 to 3
SOUTH OAK	413904087042301	413904.1	0870422.7	04/1985	Hand driven	604.46	607.37	.1	Steel	Drive point	2 to 3

Evaluation of Groundwater Levels Relative to Basements

The altitude of the water table relative to residential basement floors and the potential for high groundwater levels to cause wet basements (groundwater flooding) was evaluated by this study. This process was shown in a schematic in figure 2, diagram *B*. Groundwater-level altitudes measured at individual observation wells were compared with the elevation of a hypothetical basement floor positioned at each well. The hypothetical basement floor was assumed to be 6 ft below land surface. Areas near wells with groundwater levels that were frequently above hypothetical basement floors were at more risk of groundwater flooding. Groundwater-level hydrographs from the Beverly Shores area in this report also identify the elevation of land surface and the elevation of the base of a hypothetical 6-ft-deep basement if one were present at the well. This graphic element provides a basis to compare groundwater levels with those beneath foundations or basements and evaluate the risk of groundwater flooding during the study.

This study initially included plans to measure selected basement floor elevations to evaluate whether those were below a mapped water-table surface. Area residents were willing to permit installation of observation wells along nearby road right-of-way, but no residents gave permission to collect detailed elevation data from their individual residences; therefore, measurements of basement elevations were not made. Efforts planned by this study to survey basement elevations were redirected to installation of wells and surface-water level reference marks and to increasing the precision of the surveyed reference-mark altitudes of most wells and surface-water sites from 0.1 ft to 0.02 ft.

The widespread use of foundation underdrains and sump-dewatering systems that drain groundwater from basements, plus septic systems that recharge the surficial aquifer, also precluded the possible use of basement floor elevations to compare with the water-table altitude. Water-table measurements in basement sumps were not considered a reliable indicator of the static water level of groundwater because the sumps are frequently equipped with sump pumps that can artificially draw down the water level beneath the foundation. Wells and surface-water measuring sites used by this study were located at distance from homes to avoid artificial depression of the water table from withdrawals by sump pumps and artificial recharge of the water table by discharge from septic systems.

Relation of Hydrologic Processes to Groundwater and Surface-Water Levels and Flow Directions in a Dune-Beach Complex

Groundwater and surface-water levels fluctuate in response to complex interactions between hydrologic processes that add or remove water from the aquifer. Additions of water to the surficial aquifer result from precipitation and from wastewater derived from lake supplies to the study area, as diagrammed in figure 2. Natural groundwater outflows are routed away from the study area, either by direct seepage to the marsh and Derby Ditch, by outflow from subsurface drains to Derby Ditch, or by direct seepage to Lake Michigan. Evapotranspiration occurs when water evaporates from a shallow water table (or saturated soil surface) or water surface or is transpired as water vapor by vegetation, thereby causing a net loss of water from the aquifer.

Groundwater levels in the surficial aquifer could also be affected as surface-water levels change in discharge areas at wetlands, ditches, and Lake Michigan. A long-term rise in surface-water level at a groundwater outflow area, assuming a steady water table, could decrease the slope of the water table—the head gradient—and thereby reduce the rate at which groundwater discharges from the aquifer. A lower surface-water level at the groundwater outflow area, again assuming a steady water table, could increase the head gradient and thereby increase the rate at which groundwater discharges from the aquifer. Because some of above-mentioned conditions frequently affect the groundwater levels in the surficial aquifer and surface-water levels, their interactions can be better understood through analysis of how water levels fluctuate in the study area in response to one or more of the recharge or withdrawal processes.

Precipitation, Evapotranspiration, and Recharge to Groundwater

The years 2006-9 were the wettest 4-year period for precipitation in recent years (1952–2009; figs. 10 and 11). This period follows completion of wetland restoration in the Great Marsh (1998-2002) and the change in the source of water to Beverly Shores in 2005. Annual precipitation totals at the INDU station in 2006–9 were 50.11, 44.89, 55.75 and 43.88 in/yr respectively (fig. 10). The 2008 and 2009 annual precipitation totals were the wettest on record for the combined Ogden Dunes-INDU station record. February and March 2009 precipitation totals in the study area were from 1.5 to 2 times the monthly mean amounts for the INDU station. The 2006–9 period of data collection for this study included among the highest recorded annual, seasonal, and event precipitation totals for the study area and region (figs. 11 and 12). By comparison, 1990 was the wettest year during 1952–2009 and

1990, 1991, and 1993 were wetter than median annual precipitation for 3 of 4 years (fig. 10).

Precipitation totals from January 1, 2009, through May 1, 2009, were the second highest recorded for a station near Chicago for the period 1871 through 2009 (National Weather Service, 2009). Precipitation totals received in the region during September 12–15, 2008, from the remnants of Hurricane Ike and Tropical Storm Lowell were among the highest on record for northwestern Indiana; about 10.77 in. at the INDU station and higher totals up to 11.46 in. were reported from stations in the region (National Weather Service, 2008).

Three-year cumulative totals of precipitation in 2006–8 and 2007–9 were the highest for the 1951–2009 period of record for the combined Ogden Dunes-INDU data (fig. 11). This observation was also duplicated regionally for precipitation records from the Wanatah station and the combined record from Valparaiso (station locations in fig. 1). Cumulative 3-year

total precipitation amounts from all three stations (fig. 11) indicate that precipitation totals for 2007 through 2009 and 2006 through 2008 are the first and second highest rainfall totals on record for the region. Missing data from this record included years where one or more months of precipitation were not reported. Months with large precipitation amounts were more frequent in 2006–9. In 8 of 12 months in 2006–9, 2 or more monthly precipitation amounts at the INDU station were in the largest 25 percent of all monthly amounts for 1952–2009 (fig. 12). The months of January, February, March, July, August, September, October, and December had two to three monthly precipitation totals each from 2006 through 2009 that were greater than the 75th-percentile value. Relatively large monthly precipitation totals from July, August, and September of 2006 and 2007 were also followed by relatively high December and January (2007 and 2008) totals.

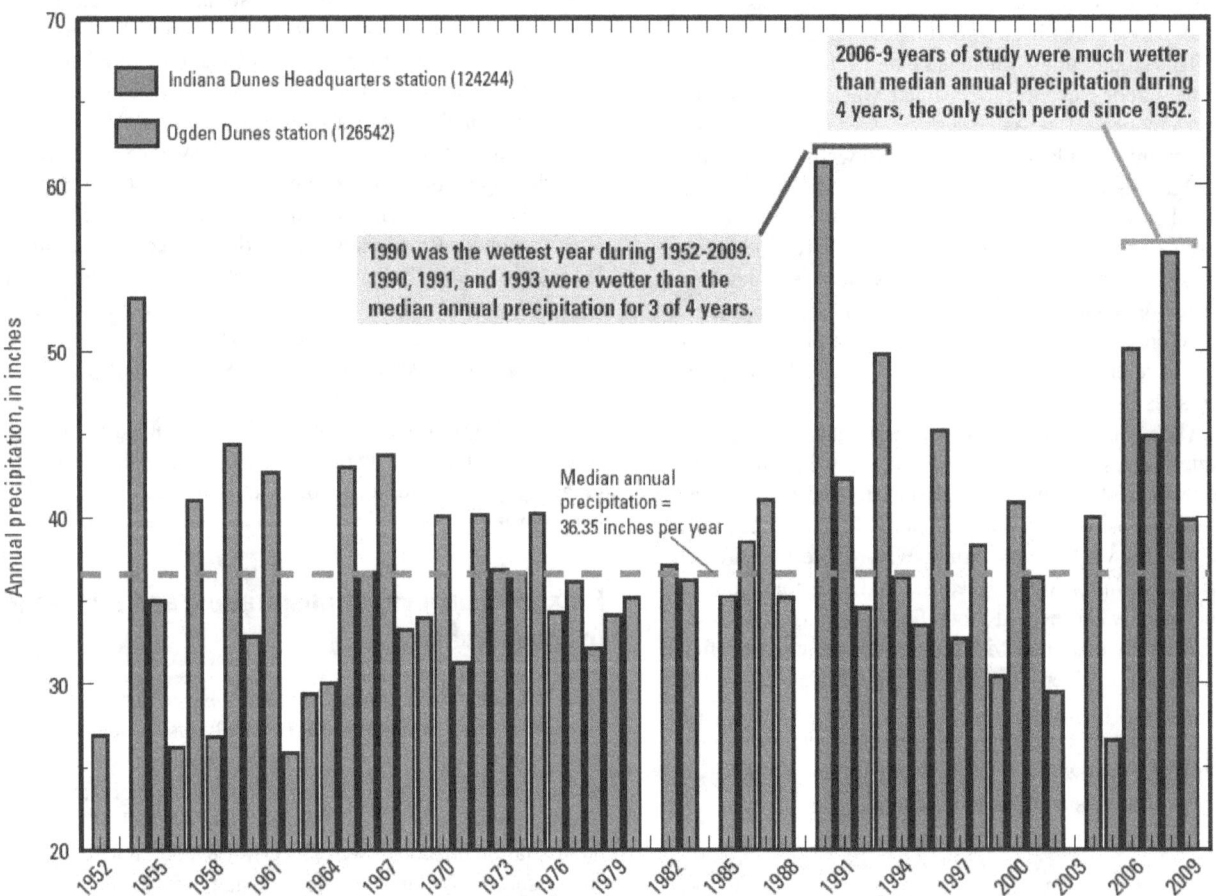

Figure 10. Annual precipitation at Ogden Dunes (1952–89) and Indiana Dunes National Lakeshore (INDU; 1989–2009), northwestern Indiana. Station locations are shown on fig. 1.

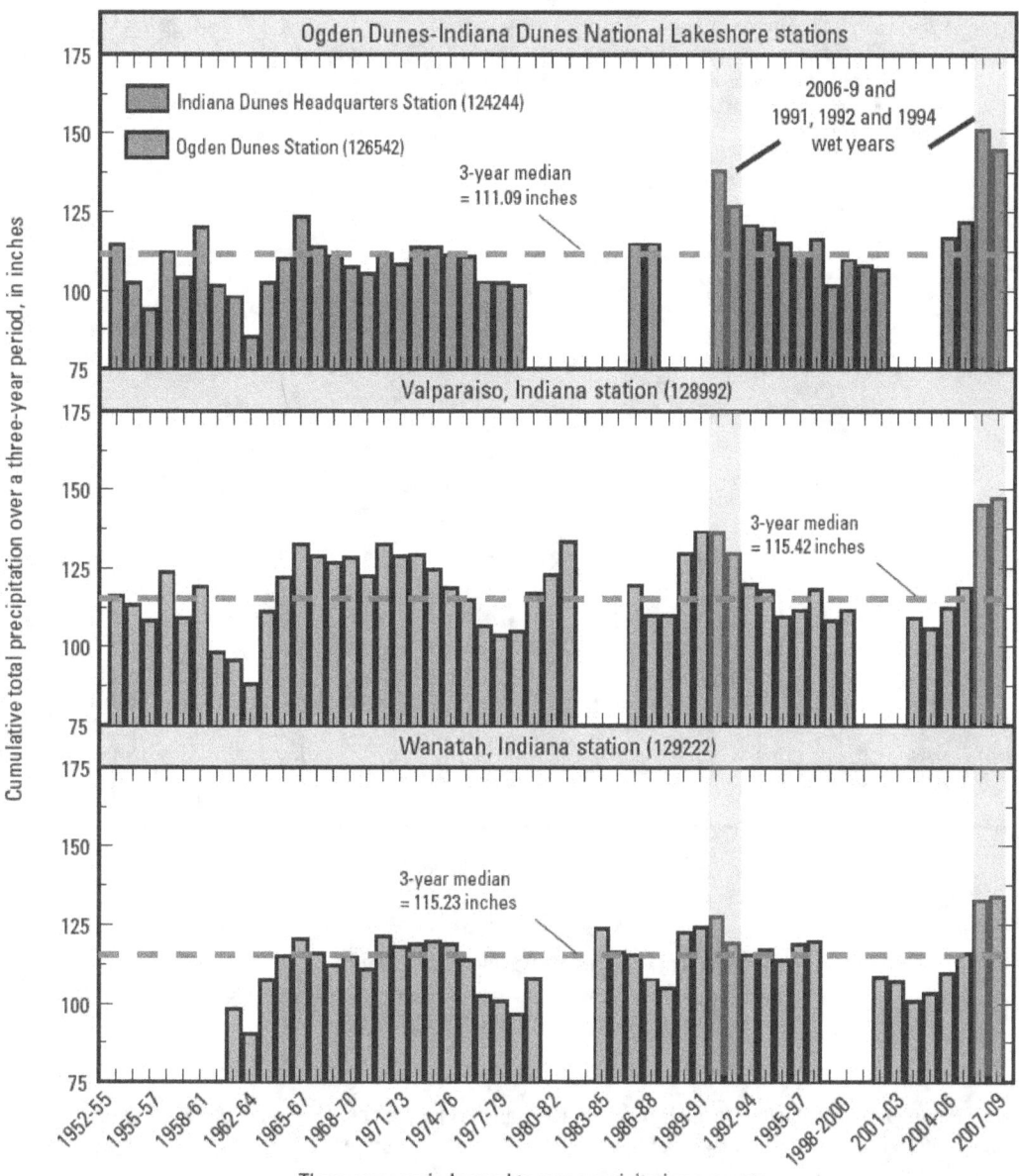

Bars on the plots were omitted when data from one or more months in the three-year period were missing.

Three-year cumulative totals of precipitation in 2006–8 and 2007–9 were the highest for the 1951–2009 period of record for the Ogden Dunes-Indiana Dunes, Wanatah, and Valparaiso stations. These years include the first and second highest annual rainfall totals on record for the region.

Figure 11.   Cumulative 3-year totals of precipitation indicating that 1990–93, 2006–8, and 2007–9 totals were regionally among the wettest 3 years in the study area and at nearby stations. Years with 1 or more missing months of record are not shown. Station locations are shown on fig. 1.

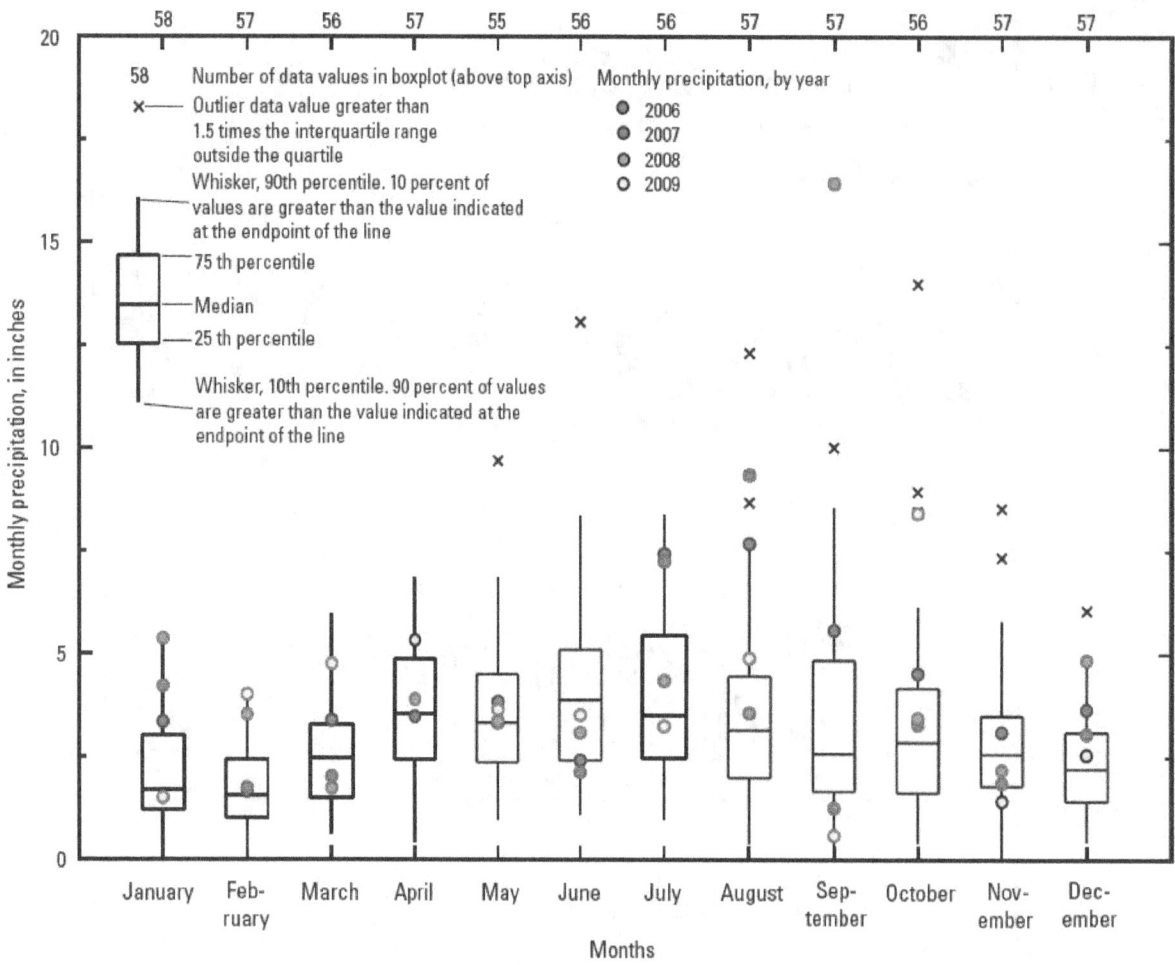

Figure 12. Box-and-whisker plots of monthly precipitation statistics for the combined records from the weather stations at Ogden Dunes (1951 through 1989) and Indiana Dunes National Lakeshore, northwestern Indiana (1989 through 2009). Station locations are shown on fig. 1.

The daily precipitation record from the INDU station was used to compare with changes in groundwater and surface-water levels at Beverly Shores and Howe's Prairie. Daily precipitation amounts from the INDU station were well correlated with and adequately represented the amounts measured at the Great Marsh station (fig. 13). The coefficient of determination (R^2 statistic) for the linear regression in figure 13 indicates that the INDU daily precipitation record can explain about 84 percent of the variability of daily precipitation at the Great Marsh station. The amount of variance explained increased to about 91 percent if the precipitation totals from a single day (July 29, 2009) were removed. On that day, daily precipitation totals were 1.34 in. at the INDU station and 2.75 in. at the Great Marsh station.

Recharge from natural sources (those excluding residential water use) was estimated to equal the difference of precipitation, minus losses from evaporation and plant transpiration (evapotranspiration), corrected for day-length and soil-moisture factors (table 6). Monthly evapotranspiration was estimated by using daily precipitation and daily mean air temperature data from the INDU station, adjusted for length of days to latitude of about 41 degrees north, using the empirical method of Thornthwaite (1948). Evapotranspiration returns more precipitation to the atmosphere as water vapor during the spring, summer, and early autumn months of May through September—the growing season—than other parts of the year (Thornthwaite, 1948; Beaty, 1994). For the first two to three months of the growing season when evapotranspiration

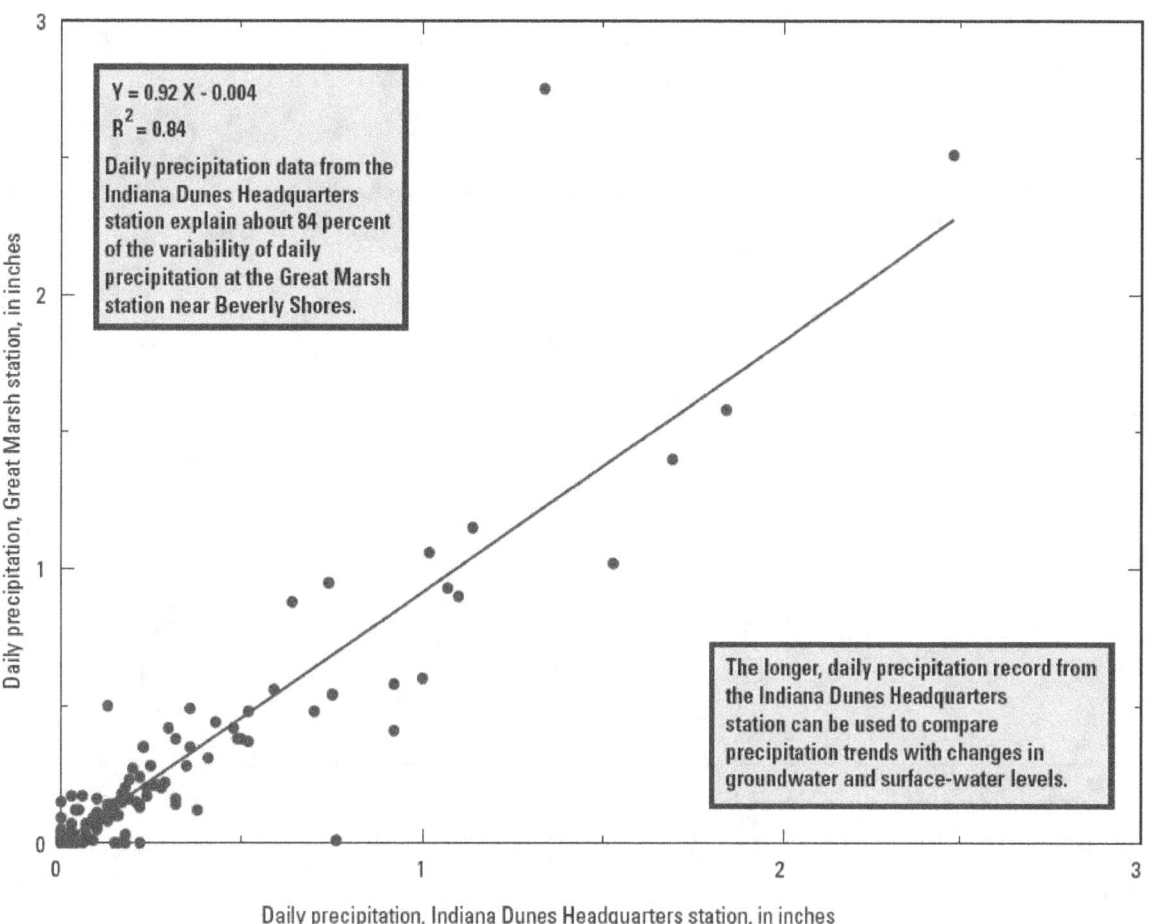

$Y = 0.92 X - 0.004$
$R^2 = 0.84$

Daily precipitation data from the Indiana Dunes Headquarters station explain about 84 percent of the variability of daily precipitation at the Great Marsh station near Beverly Shores.

The longer, daily precipitation record from the Indiana Dunes Headquarters station can be used to compare precipitation trends with changes in groundwater and surface-water levels.

Figure 13. Relation of daily precipitation amounts recorded at a station in the Great Marsh at Beverly Shores to those recorded at the Indiana Dunes National Lakeshore headquarters (station 124244), northwestern Indiana, November to December 2008 and March to December 2009. Great Marsh precipitation station is shown on fig. 4 and Indiana Dunes National Lakeshore headquarters station is shown on fig. 1.

exceeds precipitation, the part of recharge stored in the soil—the residual soil moisture capacity—was subtracted from the recharge estimate up to an assumed maximum of 2 in. per year, also by using the method of Thornthwaite (1948).

The recharge estimates computed by this study can be affected by assumptions used to compute evapotranspiration rates. The residual soil moisture capacity of 2 in. used to estimate potential evapotranspiration is similar to the available water capacity mapped for the region by Westenbroek and others (2010, fig. 18). The assumed soil moisture capacity is reasonable because the permeable, sandy soil texture under the dune-beach complex would conduct recharge more rapidly and retain less water than the overall range of clayey to sandy soils represented by Thornthwaite (1948). The recharge estimates also included an assumption that no evapotranspiration losses occur during months when the average of mean daily air temperature was at or below 32°F. The recharge estimates may potentially under represent evapotranspiration during winter months.

Recharge rates to groundwater in 2006–9 were estimated to be higher than the typical 11 in/year because annual precipitation amounts in those years were larger than the median precipitation of 36.35 in/yr (table 6). Precipitation amounts in 2006–9 were generally larger than latitude adjusted monthly evapotranspiration amounts during non-growing-season months (table 6). Estimates of recharge to groundwater from precipitation were 15.2 in/yr in 2006, 13.5 in/yr in 2007, 22 in/yr in 2008, and 20.5 in/yr in 2009 (table 6). Typical annual evapotranspiration in the Lake Michigan watershed in Indiana was estimated to be about 25 in/yr (Beaty, 1994) during median precipitation conditions of 36.35 in/yr, leaving about 11 in/yr as typical annual recharge. These recharge rates are higher than many estimates for the region (9 in/yr, Little Calumet-Galien Basin; Neff and others, 2005) and the Calumet Aquifer System (about 11 in/yr; Beaty, 1994) but are comparable to those used to simulate farther inland sand aquifers under similar hydrologic conditions in northwestern Indiana (15 in/yr; Duwelius and others, 2002) and northern

Table 6. Estimated monthly and annual recharge from precipitation and air temperature data and evapotranspiration estimates for the study area, northwestern Indiana, 2006–9.

Month and year	Monthly total precipitation, inches (P)	Mean monthly temperature, Fahrenheit	Mean monthly temperature, Celsius (T)[1]	Thornthwaite monthly heat index (i)[1] and annual heat index (I)[2] (in bold)	Thornthwaite exponent (a)[3]	Thornthwaite monthly sunlight factor, 41 degrees latitude (S)[4]	Latitude adjusted monthly evapotrans-piration, inches (ET)[4]	Estimated monthly recharge, adjusted for 2 inches residual soil moisture, inches (R)[5]
January-2006	3.35	36.1	2.3	0.3037	1.63	0.83	0.15	3.20
February-2006	1.66	30.1	-1.3	.0000	1.63	.83	.00	1.66
March-2006	3.39	39.0	3.9	.6835	1.63	1.03	.45	2.94
April-2006	3.48	53.4	11.8	3.6763	1.63	1.11	2.95	.53
May-2006	3.84	58.1	14.5	5.0117	1.63	1.25	4.64	-.80
June-2006	2.42	67.1	19.4	7.7824	1.63	1.26	7.52	-1.20
July-2006	7.44	75.0	23.9	10.6624	1.63	1.27	10.64	.00
August-2006	7.68	72.2	22.3	9.6493	1.63	1.19	8.95	.00
September-2006	5.58	62.3	16.9	6.3356	1.63	1.04	4.97	.61
October-2006	4.52	50.0	10.0	2.8560	1.63	.96	1.94	2.58
November-2006	3.1	44.7	7.3	1.7585	1.63	.82	.98	2.12
December-2006	3.65	34.8	1.6	.1713	1.63	.80	.08	3.57
Annual total 2006	**50.11**			**48.8909**				**15.21**
January-2007	4.22	29.9	-1.2	.0000	1.78	.83	.00	4.22
February-2007	1.75	19.8	-7.6	.0000	1.78	.83	.00	1.75
March-2007	2.03	43.2	6.2	1.3913	1.78	1.03	.84	1.19
April-2007	5.36	47.6	8.4	2.1744	1.78	1.11	1.54	3.82
May-2007	3.32	63.6	17.6	6.6999	1.78	1.25	6.50	-2.00
June-2007	2.13	70.5	21.5	9.1125	1.78	1.26	9.41	.00
July-2007	7.26	71.5	22.0	9.4044	1.78	1.27	9.85	.00
August-2007	9.36	73.6	23.1	10.1695	1.78	1.19	10.12	.00
September-2007	1.26	68.5	20.2	8.2944	1.78	1.04	6.96	.00
October-2007	3.28	59.7	15.4	5.4978	1.78	.96	3.96	.00
November-2007	1.86	40.5	4.9	.9721	1.78	.82	.44	1.42
December-2007	3.06	29.6	-1.3	.0000	1.78	.80	.00	3.06
Annual total 2007	**44.89**			**53.7162**				**13.46**

Table 6. Estimated monthly and annual recharge from precipitation and air temperature data and evapotranspiration estimates for the study area, northwestern Indiana, 2006–9.—Continued

Month and year	Monthly total precipitation, inches (P)	Mean monthly temperature, Fahrenheit	Mean monthly temperature, Celsius (T)[1]	Thornthwaite monthly heat index (i)[1] and annual heat index (I)[2] (in bold)	Thornthwaite exponent (a)[3]	Thornthwaite monthly sunlight factor, 41 degrees latitude (S)[4]	Latitude adjusted monthly evapotrans-piration, inches (ET)[4]	Estimated monthly recharge, adjusted for 2 inches residual soil moisture, inches (R)[5]
January-2008	5.38	25.3	-3.7	.0000	1.60	.83	.00	5.38
February-2008	3.45	24.7	-4.5	.0000	1.60	.83	.00	3.45
March-2008	1.73	34.8	1.5	.1683	1.60	1.03	.11	1.62
April-2008	3.9	50.7	10.4	3.0177	1.60	1.11	2.41	1.49
May-2008	3.37	55.5	13.1	4.2897	1.60	1.25	3.94	-.57
June-2008	3.09	70.0	21.3	8.9586	1.60	1.26	8.65	-1.43
July-2008	4.35	72.4	22.4	9.6962	1.60	1.27	9.48	.00
August-2008	3.57	70.8	21.6	9.1501	1.60	1.19	8.36	.00
September-2008	16.44	67.0	19.6	7.9065	1.60	1.04	6.26	10.18
October-2008	3.44	53.7	12.1	3.7929	1.60	.96	2.65	.79
November-2008	2.18	40.5	4.9	.9555	1.60	.82	.53	1.65
December-2008	4.85	24.9	-3.9	.0000	1.60	.80	.00	4.85
Annual total 2008	**55.75**			**47.9355**				**22.04**
January-2009	1.5	17.4	-8.1	.0000	1.53	.83	.00	1.50
February-2009	4.02	28.9	-1.2	.0000	1.53	.83	.00	4.02
March-2009	4.76	39.8	4.3	.8012	1.53	1.03	.60	4.16
April-2009	5.33	47.3	8.3	2.1380	1.53	1.11	1.74	3.59
May-2009	3.64	59.7	15.4	5.4881	1.53	1.25	5.09	-1.45
June-2009	3.52	67.6	19.9	8.0883	1.53	1.26	7.60	-.55
July-2009	3.25	68.6	20.3	8.3568	1.53	1.27	7.91	.00
August-2009	4.86	70.0	21.1	8.8639	1.53	1.19	7.87	.00
September-2009	0.59	64.0	18.1	6.9866	1.53	1.04	5.41	.00
October-2009	8.42	49.8	9.9	2.8019	1.53	.96	1.98	6.44
November-2009	1.43	46.1	8.0	2.0301	1.53	.82	1.22	.21
December-2009	2.56	28.8	-1.8	.0000	1.53	.80	.00	2.56
Annual total 2009	**43.88**			**45.5549**				**20.48**

[1] If the mean monthly temerature (t) is greater than 32 degrees Faherenheit, then the monthly heat index (i) is $i=(T/5)^{1.514}$; otherwise $I = 0$

[2] Thornthwaite annual heat index (I) is the sum of all monthly heat indexes per year (i)

[3] Thornthwaite exponent (a) is computed as $a = (0.000000675 \times I^3)+(0.0000771 \times I^2) + 0.01792 \times I + 0.49239$

[4] Estimated monthly evapotranspiration (e) is computed as $e =1.6 \times ((10 \times t)/I)^a) \times S$, where S is the Thornthwaite monthly sunlight factor for 41 degrees latitude

[5] Estimated monthly recharge (R) is computed as $R = P - ET$, except during the first two to three months when R is less than zero. For those months, the part of recharge taken up by plants and lost as evaporation, the residual soil moisture capacity (s), is subtracted from the recharge estimate up to a maximum of 2 in. per year. The residual soil moisture capacity used for the estimate was assumed to be 2 in. and is about half that assumed for typical soils by the Thornthwaite method, because of the sandy soil texture and rapid recharge under the dune-beach complex.

Indiana (24 in/yr; Bayless and Arihood, 1996). The recharge values estimated for this study area are reasonable because the sandy soils under the dune-beach complex are sufficiently permeable to allow most precipitation to infiltrate directly into the surficial aquifer.

Water-Supply Changes and Recharge to Groundwater

In 2005, residences in Beverly Shores changed their use of water from local groundwater to a Lake Michigan-derived water source (lake water source) for domestic supply. Before 2005, domestic wells in Beverly Shores withdrew groundwater from wells open to the surficial aquifer at depths less than about 60 to 70 ft below land surface and from wells in deeper, confined units (Olyphant and Harper, 1995). Residences in Beverly Shores discontinued groundwater use and were connected in summer 2005 to a lake water source. The volume of Lake Michigan water used in the study area for domestic supply increased from 2,300,900 ft³/yr in 2006 to 2,719,700 ft³/yr in 2009 (table 7). Wells in Beverly Shores were largely disconnected and closed as of 2005; most residents no longer use groundwater as a domestic water supply.

This report uses the amount of the lake water source that was supplied to the study area in 2006-9 to compute a maximum estimate of recharge from water-supply changes in the residential area of Beverly Shores in the dune-beach complex. The net effect of the change to a lake water source has been to increase recharge to the surficial aquifer. Homes in Beverly Shores use septic systems equipped with finger- or perforated-tank systems for domestic-wastewater discharge (Olyphant and Harper, 1995). Groundwater withdrawn from the surficial aquifer by residential wells originated from local precipitation and was locally recycled to the surficial aquifer at the same property.

The amount of Lake Michigan water brought into Beverly Shores may overstate the volume of water available for recharge to the surficial aquifer. Domestic water uses in the Indiana part of the Great Lakes basin have been estimated to consume about 15 percent of water supplied to residences (Shaffer and Runkle, 2007), thereby decreasing water available for discharge from the home to septic systems and for recharge to groundwater. In addition, some residences in Beverly Shores formerly withdrew water from the deeper, basal sand aquifer that was recharged outside of the study area (Buszka and others, 2007). By definition, water produced from deeper aquifers did not contribute to a lowering of the water level in the surficial aquifer. Replacement of that part of water use from the basal sand aquifer with a lake-based water source may therefore produce little or no net change in domestic water use.

It is also not known whether the water use in Beverly Shores increased after the change in water source from groundwater to surface water; and by analogy, whether the estimated recharge from the change in water source also increased because of the change. For example, the smaller iron concentrations in the lake water source relative to water from the surficial aquifer could have contributed to an increased domestic water use after the 2005 change. The median concentrations of iron reported for Lake Michigan samples near Indiana about (about 0.1 to 0.2 milligrams per liter) are less than that of groundwater from the Calumet aquifer system (about 0.9 milligrams per liter; Beaty, 1994, pages 137 and 173). The surficial aquifer is part of the Calumet aquifer system. Excessive iron in groundwater can stain clothing and plumbing fixtures, cause taste problems and clog well screens if precipitated (Beaty, 1994). The change to a lake-based water source could permit local residents to increase their use of water by diminishing the aesthetic problems associated with its use.

The net increase in recharge from the reduction of groundwater withdrawals and substitution of a lake water source was small in comparison to the recharge from precipitation to the study area. The net increase in recharge during 2006–9 from the water-source change contributed from about 13 to 40 percent of the overall increase in recharge above the typical amount of 11 in/yr. Assuming that the volume of lake water brought into to the study area was supplied over an estimated area of approximately 400 acres in Beverly Shores, the volume of lake water supplied equates to a uniform application of the net increase in recharge that ranged from about 1.58 in/yr in 2006 to about 1.87 in/yr in 2009. The percentage of total annual recharge increase in 2006–9 from the change to a lake water source ranged from about 7 percent in 2006 to 10.8 percent in 2007 (table 8).

Hydrogeology of the Study Areas at Beverly Shores and Howe's Prairie

The surficial aquifer in the Beverly Shores part of the study area is composed of dune, beach, and lacustrine sands, as described by Shedlock and others (1994) (figs. 14–15). Cores collected from the surficial aquifer during this study also included occasional small lenses of organic muck and silt that were less than 1 ft thick. The thickness of the surficial aquifer varied, ranging from about 10 ft under the Great Marsh to about 65-70 ft under a dune adjacent to Lake Michigan and thins to less than 5 ft near Lake Michigan. The saturated thickness of the surficial aquifer was about 8–10 ft, under the Great Marsh and about 16–18 ft under parts of the dune-beach complex.

A confining unit composed of till and glaciolacustrine silty clay and clay deposits underlies the surficial aquifer. Wells 554, 558, 559B, and 213G and test holes 555 and 93 were driven or drilled to the top of the confining unit; well 558 was inadvertently installed about 3-4 ft into the confining unit (well and test-hole locations in fig. 4). The thickness of the confining unit was not defined by this study. Prior drilling adjacent to well 213G described a confining-unit thickness of 107 ft. Water levels in well 558 declined very slowly after development-related

Table 7. Monthly and annual totals of Lake Michigan water supply brought into Beverly Shores, northwestern Indiana, for use by area residences, 2006–9. Water use data were available by account. One account was assumed to represent one residence

[Water use data from Michigan City Department of Water Works (Ron Plamowski, Michigan City Department of Water Works, written commun., 2009; Geof Benson, Town of Beverly Shores, written commun , 2009)]

| Month or annual total | Water use by month and year, in hundreds of cubic feet | | | | Number of accounts included in 2009 water-use data |
	2006	2007	2008	2009	
January	983	1,654	1,350	2,427	399
February	966	1,749	1,421	1,171	398
March	1,633	1,307	1,069	1,224	389
April	1,284	1,079	1,114	1,422	390
May	1,922	2,318	1,802	2,553	406
June	2,879	3,101	2,056	2,784	407
July	3,193	2,951	3,334	3,730	406
August	2,877	2,529	4,130	3,766	408
September	2,853	2,287	3,074	2,908	405
October	1,340	1,892	1,814	2,425	407
November	1,332	1,321	1,390	1,650	406
December	1,747	1,514	1,344	1,137	404
Annual total	23,009	23,702	23,898	27,197	Not computed

Table 8. Comparison of estimated annual recharge from Lake Michigan water supply, Beverly Shores, northwestern Indiana, with estimated recharge from precipitation for 2006–9 data.

Year	Annual precipitation, inches	Estimated annual recharge from precipitation (from Table 6), inches (Rp)	Estimated annual recharge from Lake Michigan water use applied over 400 acre area, inches (R(LM))	Sum of estimated annual recharge from precipitation and annual Lake Michigan water use, inches	Estimated percentage of annual recharge from Lake Michigan water supply, percent	Estimated contribution of recharge from Lake Michigan water supply greater than standard amount (11 inches/year), percent[1] (E)
2006	50.11	15.21	1.58	16.79	9.4	27.3
2007	44.89	13.46	1.63	15.09	10.8	39.9
2008	55.75	22.04	1.65	23.69	7.0	13.0
2009	43.88	20.48	1.87	22.35	8.4	16.5

[1]Estimated increase of annual recharge from imported water use (E) is computed as $E = 100*(R(LM)/(Rp - Rp(standard) + R(LM)))$ where Rp is the estimated annual recharge from precipitation, Rp(standard) is 11 inches per year and R(LM) is the estimated annual recharge from Lake Michigan water use

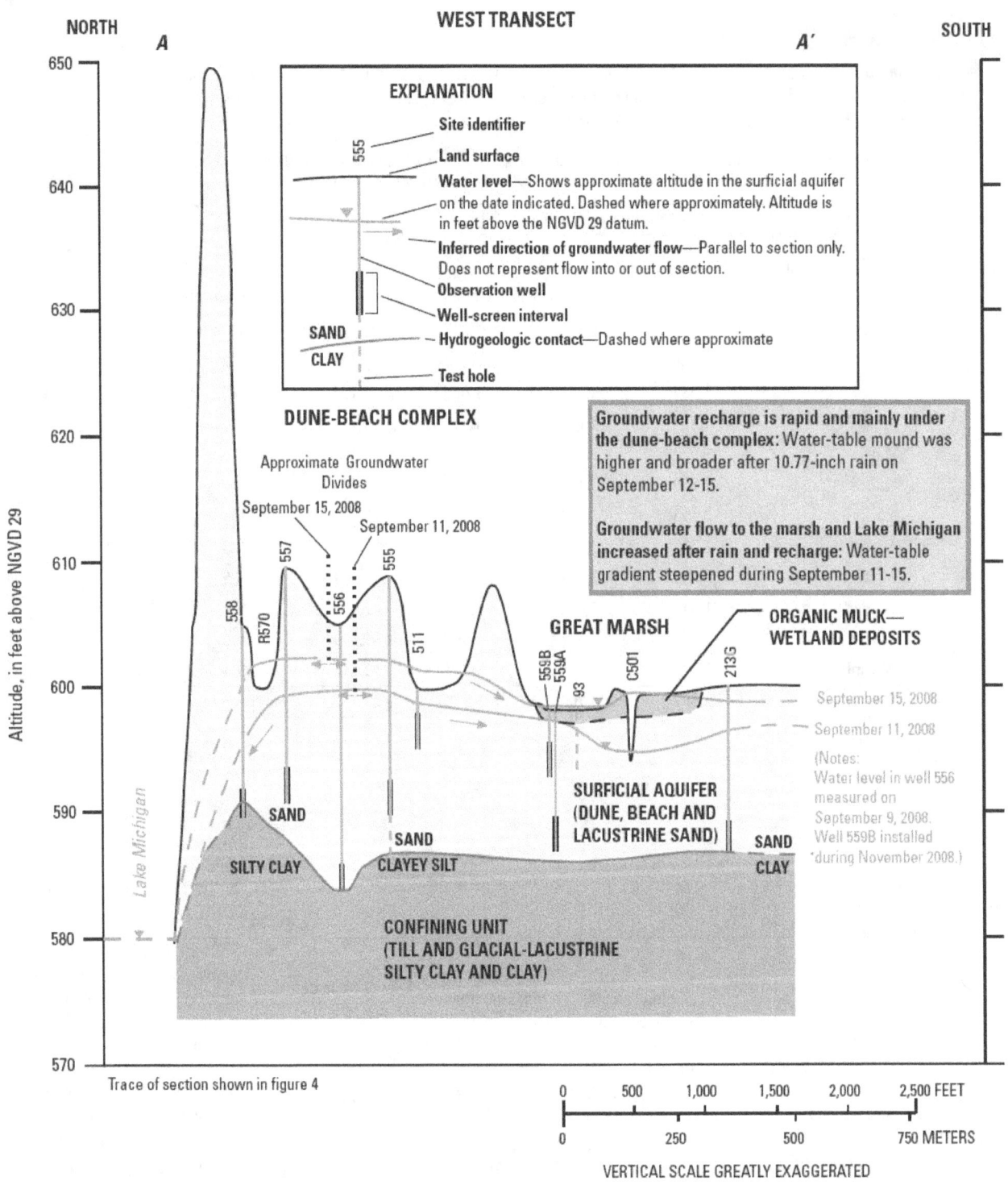

Figure 14. (above)   Hydrogeologic section *A–A′* showing the surficial aquifer and other types of sediment encountered, and approximate water-table altitudes on September 11 and 15, 2008, northwestern Indiana. Section trace is shown on fig. 4.

Figure 15. (right)   Hydrogeologic section *B–B′* showing the surficial aquifer and other types of sediment encountered, and approximate water-table altitudes on September 11 and 15, 2008, northwestern Indiana. Section trace is shown on fig. 4.

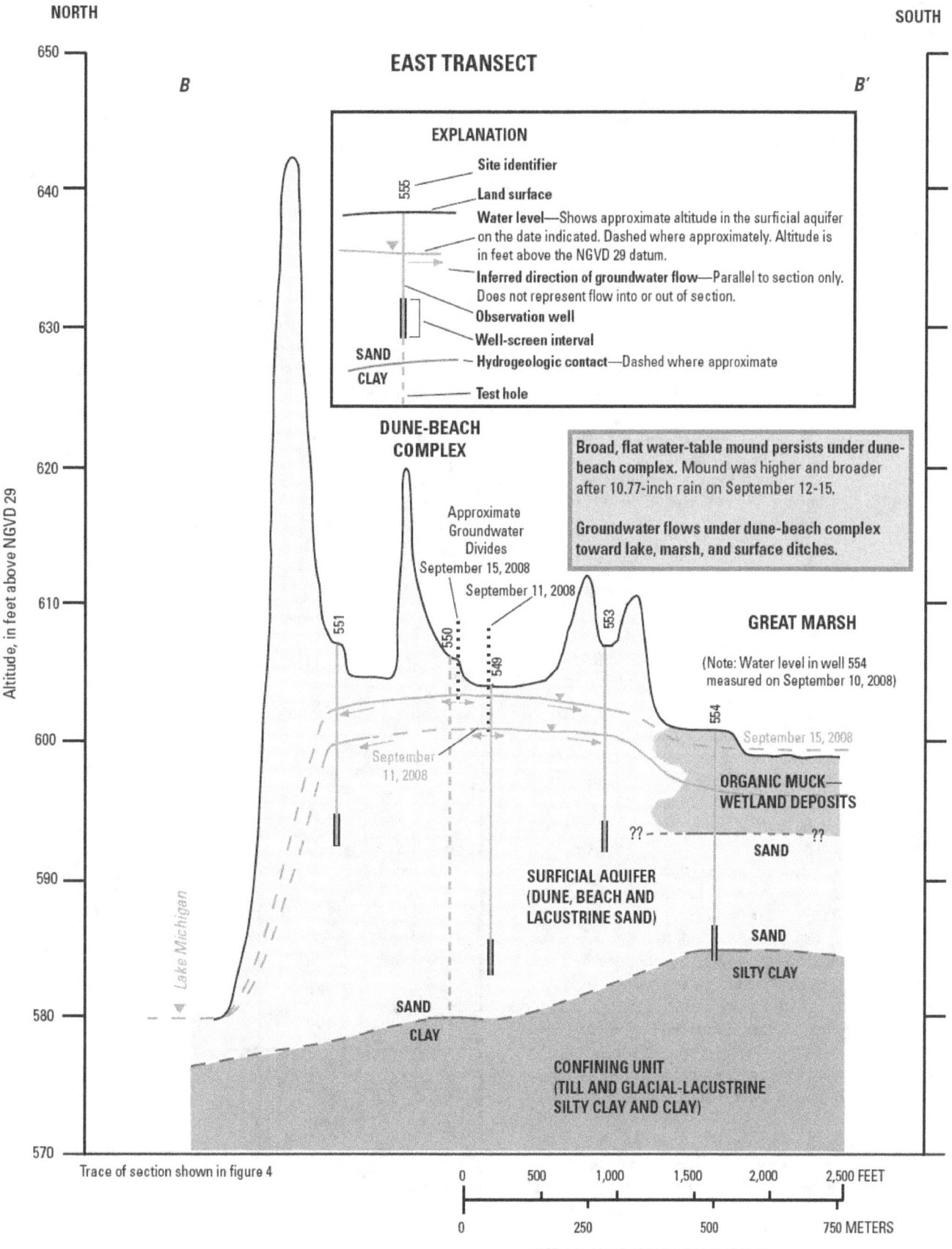

EAST TRANSECT

NORTH

SOUTH

B

B'

EXPLANATION

555 — Site identifier

Land surface

Water level—Shows approximate altitude in the surficial aquifer on the date indicated. Dashed where approximately. Altitude is in feet above the NGVD 29 datum.

Inferred direction of groundwater flow—Parallel to section only. Does not represent flow into or out of section.

Observation well

Well-screen interval

Hydrogeologic contact—Dashed where approximate

SAND
CLAY

Test hole

DUNE-BEACH COMPLEX

Broad, flat water-table mound persists under dune-beach complex. Mound was higher and broader after 10.77-inch rain on September 12-15.

Groundwater flows under dune-beach complex toward lake, marsh, and surface ditches.

Approximate Groundwater Divides
September 15, 2008
September 11, 2008

GREAT MARSH

(Note: Water level in well 554 measured on September 10, 2008)

September 15, 2008

September 11, 2008

ORGANIC MUCK— WETLAND DEPOSITS

SAND

SURFICIAL AQUIFER (DUNE, BEACH AND LACUSTRINE SAND)

SAND

SILTY CLAY

Lake Michigan

SAND
CLAY

CONFINING UNIT (TILL AND GLACIAL-LACUSTRINE SILTY CLAY AND CLAY)

Altitude, in feet above NGVD 29

650

640

630

620

610

600

590

580

570

551   550   549   553   554

Trace of section shown in figure 4

0   500   1,000   1,500   2,000   2,500 FEET

0   250   500   750 METERS

VERTICAL SCALE GREATLY EXAGGERATED

addition of water to the well; from 602.27 ft on March 13, 2009, to 597.81 ft on August 31, 2009, or about 0.29 ft/d. That observation indicates that groundwater flow rates through the confining unit are considerably less than flow rates in the surficial aquifer.

Simulations of water levels later in this report were computed using a hydraulic conductivity of 25 ft/d and a porosity of 0.3 to represent these aquifer properties. The volume and rate of groundwater flow through the surficial aquifer is directly proportional to the hydraulic conductivity of the aquifer, according to the following equations (Freeze and Cherry, 1979):

$$Q = K \times i \times A$$

$$v = K \times i / n_e$$

where

Q	is discharge, in cubic feet per second,
K	is hydraulic conductivity, in cubic feet per second,
i	is the difference in hydraulic potential (or in this case water-table altitude) between two points along a flow direction, divided by the distance between those points (dimensionless),
A	is the area across which water flows, in square feet,
v	is average linear velocity of groundwater through the aquifer, in feet per second, and
n_e	is the effective porosity (ratio of interconnected void to solid filled space in a porous media, dimensionless).

The hydraulic conductivity value of 25 ft/d that was used to represent surficial aquifer characteristics was within the range of values used by prior investigations of the aquifer. Horizontal hydraulic conductivity of the surficial aquifer ranged from about 18 to 45 ft/d (Watson and others, 2002), with a median value of about 28 ft/d. Lower hydraulic conductivities (8 ft/d) were calculated for an aquifer interval that contained relatively more organic matter than other intervals (Watson and others, 2002). These values were calculated from single-well tests of wells open to the aquifer about 1.3 mi southwest of the study area. Although single-well tests can sometimes yield lower hydraulic conductivities than multiple-well-based methods because of local well characteristics, the surficial aquifer sediments are sufficiently permeable to minimize those differences in practice. Other authors have reported values of hydraulic conductivity for other parts of the surficial aquifer that are close to the upper range of those reported by Watson (Shedlock and Harkness, 1984, 50 ft/d; Shedlock and others, 1994, 40 ft/d).

This study uses a porosity of 0.3 to represent the freely draining "effective" porosity in the surficial aquifer. The volume of the aquifer is composed of sediment grains and pore space between the grains; the fraction of pore space in a volume of aquifer is its porosity. The porosity of the surficial aquifer, measured as volumetric water content, was about 0.4

at a site in Beverly Shores (Olyphant and Harper, 1995), 0.34 at about 8.5 mi west-southwest of the study area (Isiorho and others, 1994) and was assumed to be 0.3 for two other studies in the area (Shedlock and Harkness, 1984; Watson and others, 2002). In addition, the volumetric water content of the unsaturated zone changes with time as recharge pulses infiltrate to the water table. After a recharge event infiltrates through the unsaturated zone, part of the porosity may be filled with water that surrounds the sediment grains by surface tension or that is trapped in poorly-drained pores (fig. 16). After a several day period of little or no precipitation, Olyphant and Harper (1995, p. 14) reported a volumetric moisture content of about 0.07 to 0.1 in the unsaturated zone at a site in Beverly Shores on July 30, 1995. As an approximation, this report uses an effective porosity value of 0.3, the total porosity of 0.4 minus the fraction of water remaining in the unsaturated zone of 0.1 to represent the part of the aquifer that freely drains or saturates by gravity (fig. 16).

In the Great Marsh, the core near well 554 on section *B–B′* and the hand-driven well 559B on section *A–A′* both included organic sediments (figs. 14–15). The thickness of organic sediments at well 559B could not be confirmed; an organic muck was detected as a smeared layer over the well screen during a first attempt at well installation at the site. On the basis of approximately 15 cores obtained in and near the Great Marsh in the study area during the 1980s, total organic

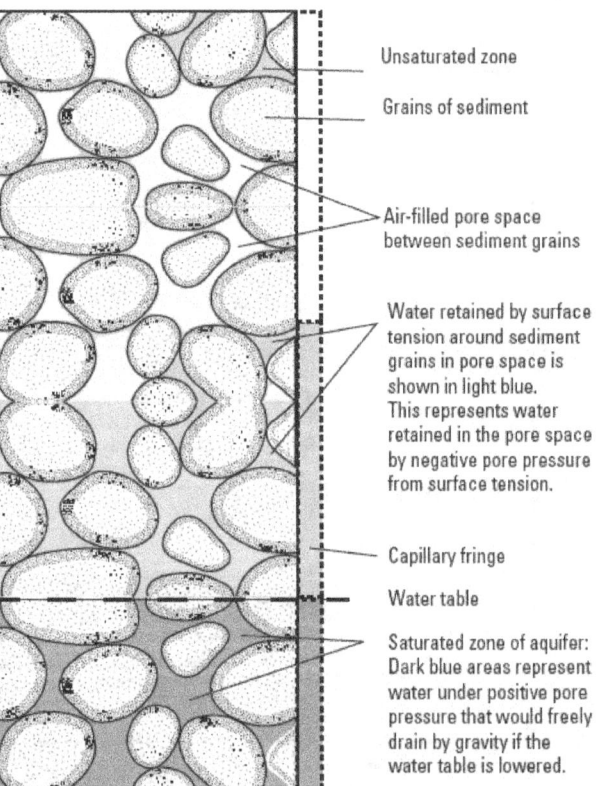

Figure 16. Conceptual diagram of the distribution of saturated porosity through the unsaturated zone, capillary fringe and saturated zone.

sediments, composed mostly of vegetable matter, muck, and some marl, ranged in thickness from less than 1 ft to as much as 10 ft (Todd Thompson, Indiana Geological Survey, written commun., 1989, 1992). Water levels in well 559B recovered to within 0.5 ft of original level more slowly—by about 2 hours after pumping for development—than other wells in the surficial aquifer installed for this study, indicating that the sand and muck deposits at this site were less transmissive of water than the surficial aquifer sand. Water levels in other wells in the surficial aquifer had water levels that fully recovered within about 5 to 15 minutes after development pumping.

The sections also illustrate that groundwater recharge is rapid and mainly under the dune-beach complex. The water-table mound was higher and broader on September 15 than on September 11 because of recharge from infiltrating precipitation (figs. 14–15). Groundwater flow velocities and volumes to the marsh and lake also increased after the September 12–15 precipitation, as indicated by the steeper water-table gradient from the dune-beach complex to the marsh on September 15 (figs. 14–15). Groundwater flow directions on sections A–A' and B–B' are schematic; their accuracy is limited because water-table contours on those days indicate flow directions that were not parallel to the sections. More accurate flow directions are indicated on water-table maps later in this report. The flow directions shown by the sections are accurate in that groundwater flows from a groundwater divide under the dune-beach complex to discharges at the Great Marsh, Lake Michigan, and adjacent ditches.

Lake Michigan was identified in prior studies as an area of low water levels where groundwater flow discharges from the surficial aquifer to the lake (Shedlock and others, 1994; Greeman, 1995). The water level of Lake Michigan used to map water-table altitudes and interpret groundwater-flow directions in figures in this report was assigned to be 580 ft. This value is within the range of daily mean water levels from June 1985 to September 1992 (from 578.58 to 583.10 ft; Greeman, 1995, table 3, site S-23) but is slightly greater than the range of mean monthly water levels during 2007–9 (575.35 ft in December 2007 to 579.91 ft in June 2009; National Oceanic and Atmospheric Administration, 2010). Lake Michigan water levels were recorded at a site about 30 mi west of the study area.

The Howe's Prairie part of the study area has a similar hydrogeology to that of the Beverly Shores part, as described by Shedlock and others (1994). A surficial aquifer underlies Howe's Prairie. The surficial aquifer is also composed of dune, beach, and lacustrine sands, has a similar saturated thickness of about 10–20 ft, and is underlain by a confining unit of till and glaciolacustrine silty clay and clay deposits (Shedlock and others, 1994, p. 22, 24–25, and 55). Regional groundwater flow directions through the area are generally mapped as flowing southward from a groundwater divide toward the Great Marsh and northward from the divide toward Lake Michigan (Shedlock and others, 1994, p. 34–35). That groundwater divide is interpreted to be near well Wet Prairie (fig. 5), on the basis of water-level altitudes discussed later in this report.

Water-Level Fluctuations and Flow Directions

Groundwater and surface-water levels fluctuate in response to complex interactions between hydrologic processes that add or remove water from the aquifer. Wells and surface-water-level measurement points and the context of their data in evaluating these interactions are described in figures 17A and 17B. The additive and withdrawal processes that most affect water-level fluctuations in the surficial aquifer are discussed in this section relative to these questions:

- How do water levels compare from periods before wetland restoration (1978–89) and afterward (2007–9) in the Great Marsh at Beverly Shores?

- How did groundwater levels change in a similar nearby but undeveloped, natural wetland area (Howe's Prairie) during the same period?

- Do groundwater levels in the dune-beach complex indicate a potential for groundwater flooding of basements?

- What are the directions of groundwater flow relative to the Great Marsh and Lake Michigan, how do they change before and after precipitation, and what do the flow directions indicate about processes that affect groundwater levels?

- Which natural and human-affected hydrologic processes most affect groundwater and surface-water levels and their short-term (daily-monthly) and longer term (seasonal) fluctuation in the dune-beach complex? These processes include:

  - recharge from precipitation (additive),

  - decreased groundwater withdrawal from the change in water supply (additive)

  - recharge of the surficial aquifer by seepage from the restored wetland (additive)

  - groundwater discharge from the surficial aquifer to the restored wetland (removal)

  - evapotranspiration (removal),

  - other drainage, such as tile drains (removal).

The last question is addressed through data analysis and through a simulation of the relative importance of these processes by using an analytical solution to render a hypothetical simulation of the surficial aquifer under the dune beach complex. Processes compared with the analytical solution were the effects of recharge from precipitation, recharge from water-supply changes, and a change in post-restoration wetland water levels.

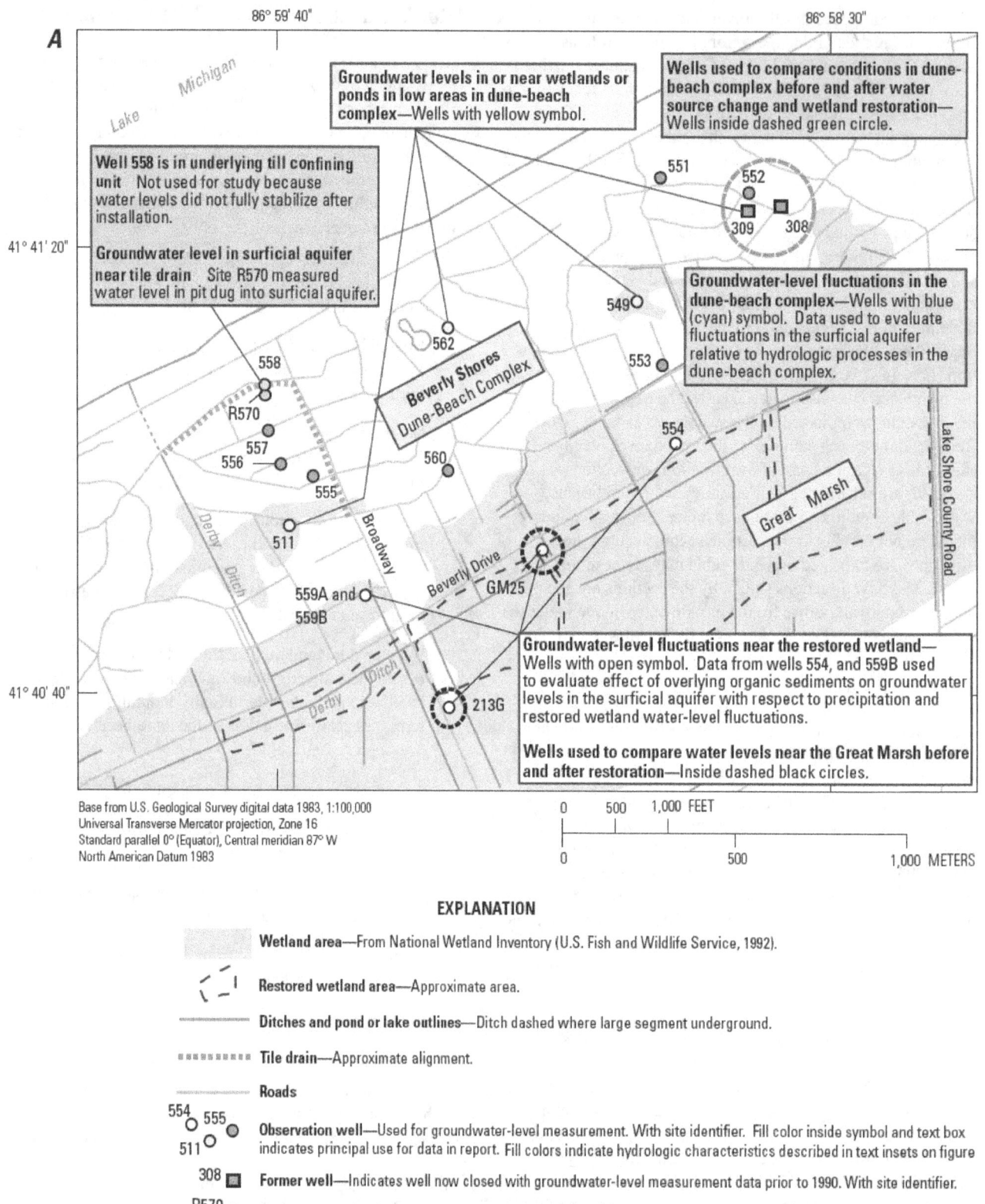

Figure 17. Wells and surface-water-level measurement points and their context in the analysis of hydrologic interactions on groundwater levels near the dune-beach complex and restored wetland area at Beverly Shores, Indiana. (*A*) Wells and groundwater levels. (*B*) Surface-water-level measurement points and surface-water levels.

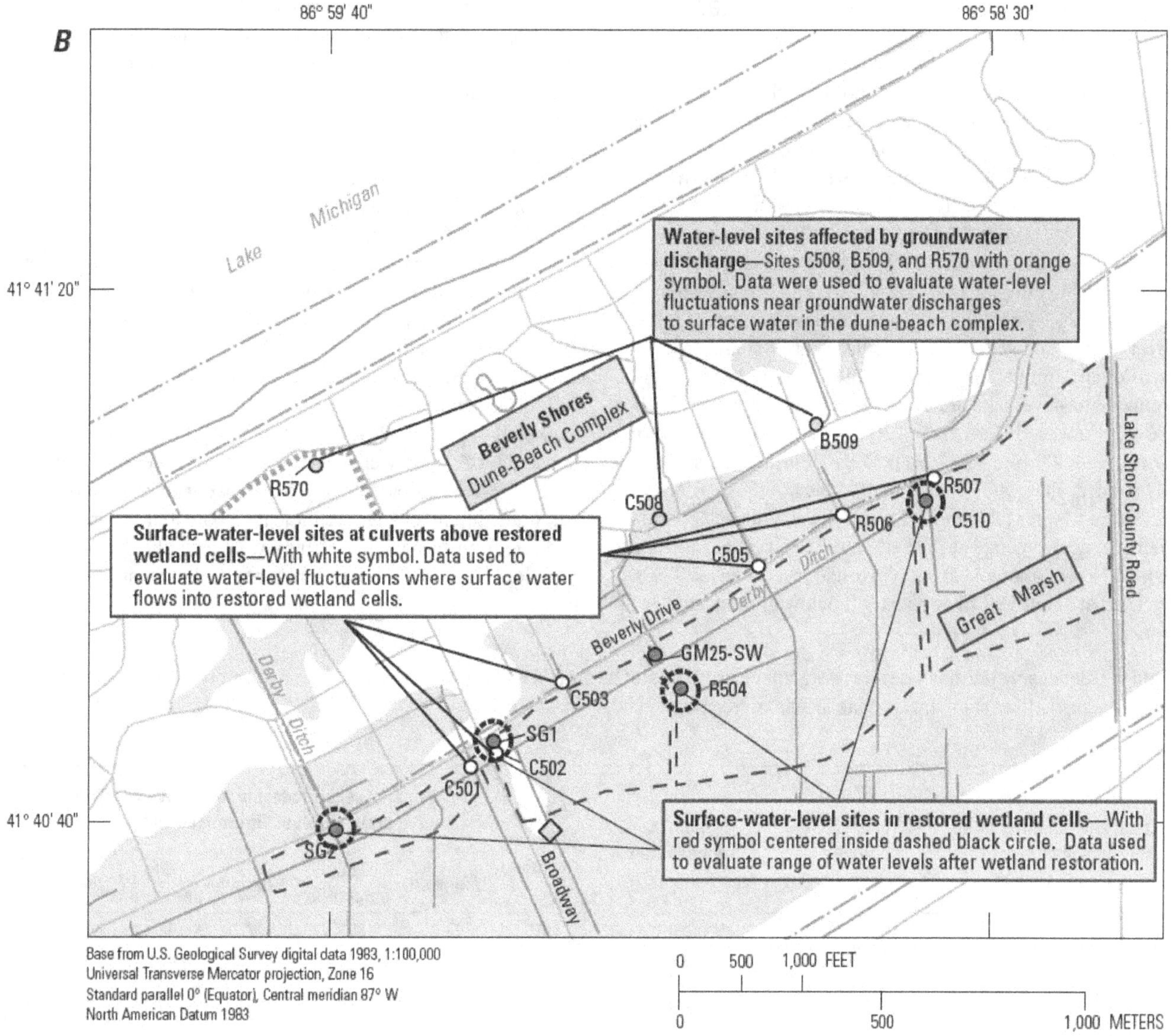

B

86° 59' 40"   86° 58' 30'

41° 41' 20"

Lake Michigan

Water-level sites affected by groundwater discharge—Sites C508, B509, and R570 with orange symbol. Data were used to evaluate water-level fluctuations near groundwater discharges to surface water in the dune-beach complex.

Beverly Shores Dune-Beach Complex

B509

R570

Surface-water-level sites at culverts above restored wetland cells—With white symbol. Data used to evaluate water-level fluctuations where surface water flows into restored wetland cells.

C508   R507
          R506   C510
     C505

Derby Ditch

Great Marsh

Lake Shore County Road

Beverly Drive

Derby Ditch

GM25-SW
C503   R504

SG1
C502
C501

Surface-water-level sites in restored wetland cells—With red symbol centered inside dashed black circle. Data used to evaluate range of water levels after wetland restoration.

41° 40' 40"

SG2

Broadway

Base from U.S. Geological Survey digital data 1983, 1:100,000
Universal Transverse Mercator projection, Zone 16
Standard parallel 0° (Equator), Central meridian 87° W
North American Datum 1983

0   500   1,000 FEET

0   500   1,000 METERS

EXPLANATION

Wetland area—From National Wetland Inventory (U.S. Fish and Wildlife Service, 1992).

Restored wetland area—Approximate area.

Ditches and pond or lake outlines—Ditch dashed where large segment underground.

Tile drain—Approximate alignment.

Roads

Indiana Dunes National Lakeshore—Approximate boundary.

R504 C508   **Surface-water-level measurement points**—With site identifiers. Fill colors indicate hydrologic
      C505   characteristics described in text insets on figure

Precipitation measurement site—U.S. Geological Survey, Great Marsh precipitation station. Temporary station maintained in 2008-9.

Water-Level Fluctuations Before and After Wetland Restoration at Beverly Shores

A review of historical data identified wells in three locations where water-level data could be compared between pre-wetland-restoration and post-wetland-restoration conditions. Records for pre-restoration groundwater levels for the water table in the study area were available for nine sites (GM23, GM24, GM25, GM26, GM27, GM28, 213G, 308, and 309). Among these sites, current (2007–9) groundwater-level data were available only for GM25 and 213G (fig. 17A). Wells GM25 and 213G are adjacent to restored wetland areas in the Great Marsh, and water-level measurements were recorded occasionally from about 1978 to 1989. Current (2007–9) groundwater-level data are available at well 552, which is about 360 ft northwest of former well 308 and about 150 ft north of former well 309 (fig. 17A). Well 552 and former wells 308 and 309 are near the same general point in the flow system; they are within a water table mound under the dune-beach complex. Well 308 was screened at the bottom of the surficial aquifer at about 594 ft or about 64 ft below ground level. Well 309 was screened at about 598 ft, or about 9 ft below ground level.

The history of groundwater-level fluctuations near a natural part of the Great Marsh at Howe's Prairie was evaluated by using water levels measured in five wells from 1985 to 2005, then again from April to September 2009 (fig. 5, in "Description of Study Area" section). Four of these wells were named previously according to local ecological or environmental characteristics: Wet Prairie, Pin Oak, North Oak, and South Oak (fig. 5).

A comparison of pre- and post-restoration water levels indicate that groundwater and surface-water levels were about 1.1 ft higher after restoration behind the wetland control structure near R504 but only slightly higher after restoration behind the wetland control structure near SG1 (fig. 17B, table 9). In comparison, water levels at a well (552) in the dune-beach complex after restoration were similar to those measured in prior wells (308 and 309) before wetland restoration and water-supply changes. In addition, water levels in well 309 were sufficiently high before wetland restoration and water-supply changes to indicate a potential for basement flooding from seepage in nearby low areas in the dune-beach complex. Details of these observations and comparisons follow.

Table 9. Summary statistics for surface-water and groundwater levels (1979–2009) measured at sites near restored parts of the Great Marsh, northwestern Indiana. Site locations are shown in figs. 17A and 17B.

[NGVD29, National Geodetic Vertical Datum of 1929]

Local site identifier or description of data sources	Years of data record	Months represented	Number of water levels	Water-level altitudes, in feet above NGVD29 or water-level differences, in feet				
				Minimum	25th percentile	Median	75th percentile	Maximum
C502	1984 to 1989	All	12	593.75	593.94	594.24	594.42	594.74
	2007 to 2009	All	162	594.52	594.91	595.08	595.24	599.34
Staff Gage SG1	2008 to 2009	All	66	596.75	597.01	597.10	597.20	599.40
Difference between paired SG1 and C502 water levels on same day	2008 to 2009	All	66	0.06	1.97	2.02	2.07	2.91
Well 213G	1979 to 1989	All	19	593.20	593.98	596.82	597.08	597.32
		October to April	10	596.14	596.85	597.01	597.24	597.32
		May to September	9	593.20	593.41	593.66	596.49	597.30
	2007 to 2009	All	161	592.09	596.71	596.94	597.16	598.81
		October to April	41	592.09	596.78	596.85	597.02	598.08
		May to September	120	595.27	596.71	597.00	597.17	597.30
Well GM 25	1979 to 1989	All	22	594.32	595.87	596.67	596.83	597.12
		October to April	12	595.93	596.76	596.82	596.96	597.12
		May to September	10	594.32	595.18	595.69	596.32	596.83
	2007 to 2009	All	158	596.65	597.27	597.38	597.49	599.34
		October to April	45	597.21	597.35	597.41	597.51	598.83
		May to September	113	596.65	597.26	597.36	597.47	599.34
R504	2007 to 2009	All	134	596.80	597.81	597.94	598.17	599.39

Median groundwater levels from well GM25 in 2007–9 after wetland restoration were about 0.6 ft higher than the median seasonal high groundwater levels in that well in October–April data from 1979–1989 (table 9, fig. 18). Groundwater levels were typically highest in well GM25 during October–April; those months are when evapotranspiration is lowest and recharge from precipitation and snowmelt is highest. Well GM25 is on the west side of a roadbed between two cells of a constructed wetland; water levels in those cells were measured at sites R504 (upstream) and SG1 (downstream). Groundwater levels before wetland restoration in well GM25 during 1979–1989 were more variable than those measured post-restoration in 2007–9. Minimum water levels in GM25 during 1979–89 were about 2.8 ft lower than maximum water levels. Minimum water levels in GM25 during 2007–9 were about 2.6 ft lower than maximum water levels. The range of post-restoration water levels was similar to pre-restoration water levels because of two extreme precipitation events in 2006–9.

Seasonally flooded conditions comparable to current water levels were indicated in pre-restoration groundwater levels for the part of the marsh immediately east of Broadway before restoration. Pre-restoration, seasonal high groundwater levels in well 213G near Broadway during October to April 1979–89 (597.01 ft) were similar to surface-water levels in the restored wetland in 2007–9 at surface-water measuring point SG1 (597.10 ft; table 9). Well 213G is next to part of the restored wetland immediately east of Broadway (fig. 17A). The median groundwater level in well 213G before wetland restoration (1978–89, 596.82 ft) was also similar to the median value after restoration (2007–9, 596.94 ft; table 9). These median groundwater levels were both similar to post-restoration median surface-water levels in the adjacent wetland at site SG1 (597.1 ft; table 9). Pre-restoration groundwater levels at well 213G that were at or slightly above the median value were measured in November to May 1979–89 (fig. 18). Locally common hydrologic conditions that have temporarily increased surface-water and groundwater levels in and near the Great Marsh before and after restoration included partial blockage of drainage culverts by vegetation and flow reduction behind small beaver dams.

Seasonal high water-level conditions before wetland restoration are represented by the subset of October to April 1979–89 water levels at GM25, 213G, and other wells with pre-restoration water levels. The seasonal high water-level conditions are important because they represent the condition when the pre-restoration wetland was seasonally flooded. The pre-restoration water levels in 1979–89 were measured during a period when annual precipitation amounts were close to the median annual precipitation; they ranged from about 5 in/yr below to 5 in/yr above median annual precipitation (fig. 10). These conditions were drier than prevailed in 2007–9.

The difference of 1.1 ft between median surface-water levels in the restored wetland at surface-water measuring point (also called "site") R504 in 2007–9 and seasonal high, pre-restoration groundwater levels in nearby well GM25 was used later in this report to represent pre- and post-restoration surface-water-level differences in the wetland. The median surface-water level at site R504 in the restored wetland cell was about was about 1.1 ft higher than the median of seasonal high groundwater levels at well GM25 (October to April 1979–89; table 9). In comparison, the difference was slightly smaller—about 0.8 ft—between the medians of post-restoration higher surface-water levels in the upstream wetland cell at site R504 and lower water levels in the downstream wetland cell at SG1 (table 9).

The groundwater levels occasionally measured (about 2–3 times per year) at wells GM25 and 213G in the surficial aquifer adjacent to the Great Marsh during 1978–89 may not represent the full range of water levels at these wells. These wells were measured on a greater frequency than other wells—at first intermittently in 2007–8, then weekly during 2008–9—and well after restoration of wetlands in the Great Marsh was completed in 2002 (fig. 18).

Groundwater levels near well 552 in 2008–9—within the dune-beach complex—are similar to those in historical data from nearby prior wells 308 and 309 in 1982–89 before water-supply and wetland-restoration changes. Historical (1982–89) water levels in wells 308 and 309 are within less than 1 ft of each other with only one exception, and they generally range from 599.6 to 603.8 ft (fig. 19). Current water levels in well 552 ranged from 600.50 to 603.54 ft—these ranges are similar to the ranges of water levels in wells 308 and 309 during 1982, 1985, and 1986 (fig. 19).

Water levels in well 309 indicated a potential for basement flooding from seepage in nearby low areas in the dune-beach complex before wetland restoration and water-supply changes. Water levels in well 309 through the 1981–88 period of record were consistently above the base of a 6-ft-deep hypothetical basement, even during a year of relatively low precipitation in 1988. The well was in a low area between dune ridges in the dune-beach complex. Sites near the former well are now developed with homes, although it is not known whether they have basements.

Surface-water levels in restored wetlands in the Great Marsh west of Broadway after wetland restoration and after the higher precipitation amounts in 2007–9 were higher than in 1984–9. For example, the range of median water levels just upstream from the restored part of the Great Marsh from Broadway to about 1,500 ft west-southwest, as reflected in data from site 502, were about 0.8 ft higher in 2007–9 after restoration than in 12 measurements made from 1984 to 1989 before restoration (table 9). Earthen berms used in the restored wetlands are designed to impound water and control surface-water runoff so that water levels are more stable and do not decline as rapidly during seasonal dry conditions. Water levels at site 502 may also have been more variable in 2007–9 because of vegetation in the culvert, sustained flows from the upstream wetland cell, or the higher amounts of precipitation falling in 2007–9 as compared to the drier 1984–89 period (fig. 11).

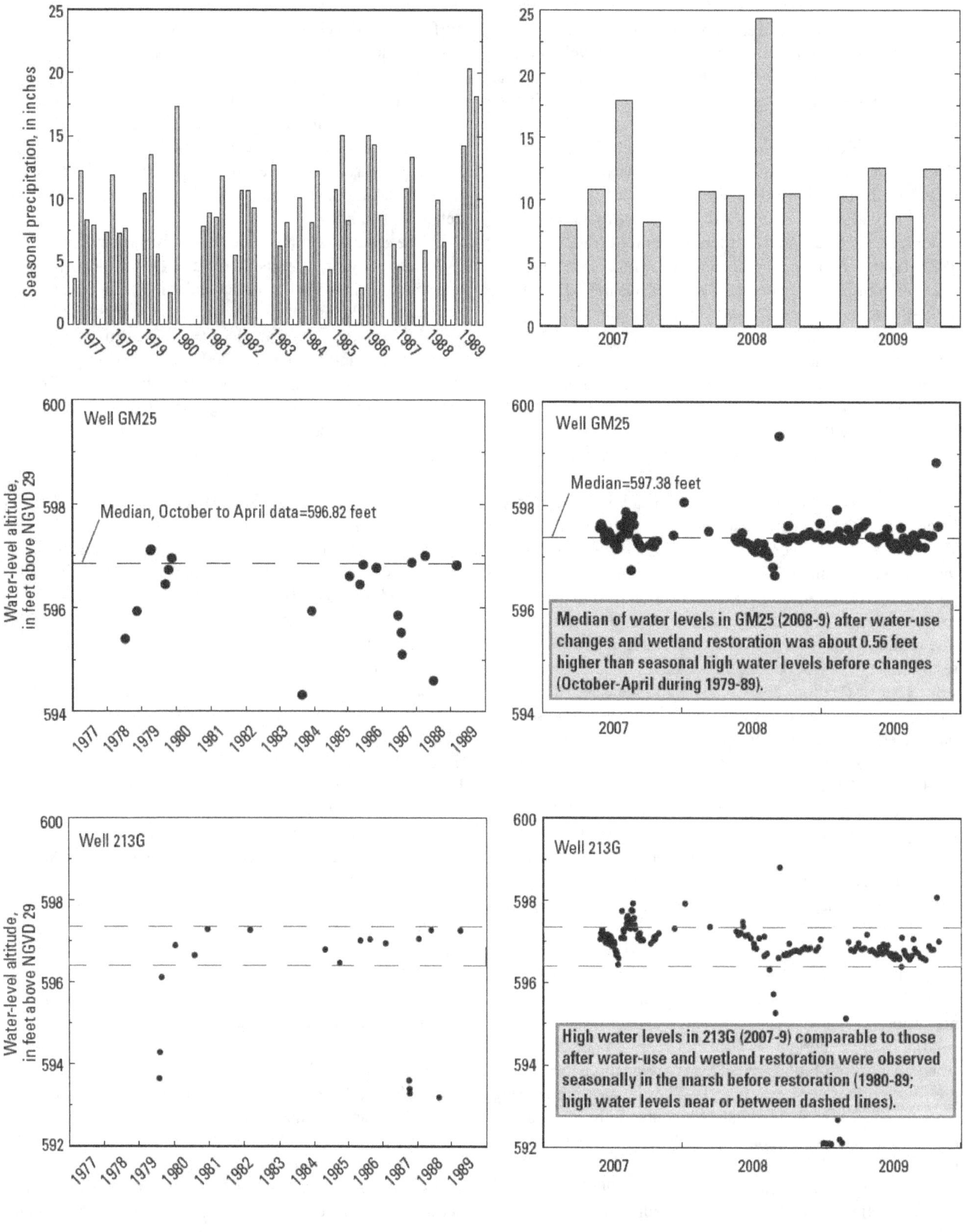

Figure 18.   Precipitation and comparison of pre-restoration (1978–89) and post-restoration (2007–9) groundwater levels at wells GM25 and 213G near a restored wetland in the Great Marsh, northwestern Indiana. Well locations are shown on fig. 17*A* and precipitation gage location is shown on fig. 1

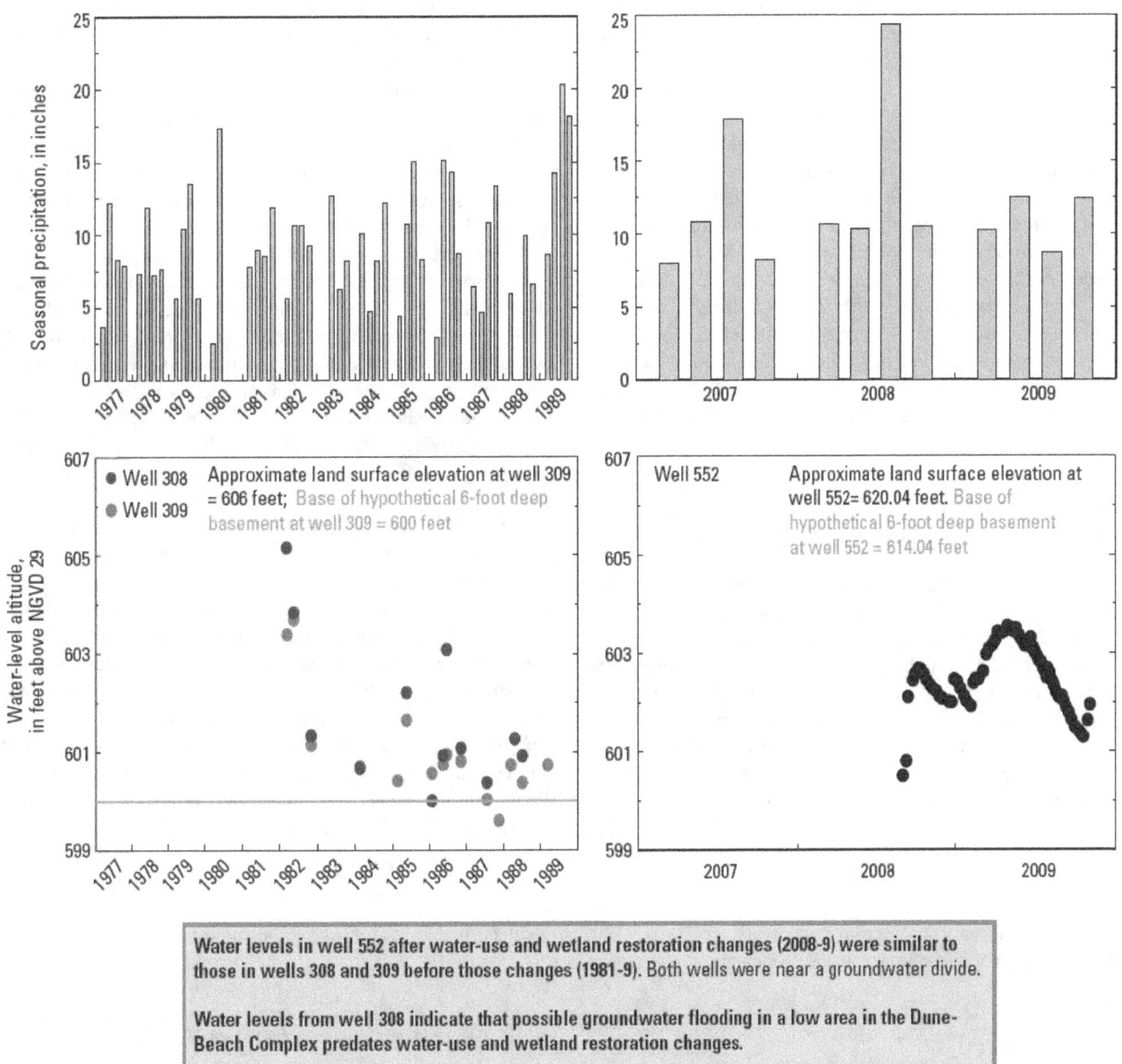

Water levels in well 552 after water-use and wetland restoration changes (2008-9) were similar to those in wells 308 and 309 before those changes (1981-9). Both wells were near a groundwater divide.

Water levels from well 308 indicate that possible groundwater flooding in a low area in the Dune-Beach Complex predates water-use and wetland restoration changes.

Figure 19.   Precipitation and comparison of pre-restoration (1978–89) groundwater levels at wells 308 and 309 and post-restoration (2007–9) groundwater levels at well 552 in the dune-beach complex, northwestern Indiana. Well locations are shown on fig. 17A and precipitation gage location is shown on fig. 1.

Median water levels in an area of pooled water behind a wetland restoration control in the Great Marsh at staff gage SG1 (table 9) averaged about 1.1 ft higher than the land surface. Land surface at SG1 is the submerged wetland soil surface at the base of the staff gage in the restored wetland cell near the southeast corner of Beverly and Broadway. Water levels in the restored wetland cell above SG1 reflect the stable water level in the restored wetland cell that extends eastward about 1,400 ft from SG1 to GM25 and southward about 750 ft from SG1 to just north of well 213G. Median surface-water levels were about 2 ft higher at SG1 in the restored wetland cell than at downstream site C502; the difference represents a locally raised water level behind the wetland restoration control. As stated earlier, water levels in 2007-9 at wells GM 25 and 213G were about 1 ft higher than those measured wells during wetter parts of the years 1979-89 before wetland restoration.

Groundwater-Level Fluctuations Without Wetland Restoration at Howe's Prairie

High groundwater-level altitudes at Howe's Prairie in 2009 near a natural wetland were caused by infiltration of the high precipitation amounts received by the area from December 2008 through April 2009. Groundwater levels from Howe's Prairie in 2009 were similar to the highest water levels measured during prior wet periods (table 10, figs. 20–21, well locations on fig. 5 in "Description of Study Area" section). For example, high groundwater levels similar to those measured in 2009 also were measured in the late summer and autumn months of 1990 in the five Howe's Prairie wells (figs. 20–21);

these measurements followed a 7.77-in. rainfall on August 18, 1990, and other relatively high rainfalls into autumn 1990. Similarly, high water levels approximately equivalent to those in 2009 from the Howe's Prairie wells also were measured during spring months in 1991 and 1993. Annual precipitation amounts in 1990 were the highest on record and in 1993 were the fourth highest on record (fig. 10), an indication that infiltrating precipitation or snowmelt was the cause of high groundwater levels at Howe's Prairie during those years. The high groundwater levels were measured in a part of the dune-beach complex next to a natural part of the Great Marsh, with no human-affected inputs to the wells. The Howe's Prairie groundwater-level data also indicate that recharge from similarly high precipitation amounts was a likely cause of high groundwater levels in other parts of the dune-beach complex, such as at Beverly Shores.

Water levels in the Howe's Prairie wells typically rose during non-growing-season months (October–March; shaded gray bars in figs. 20–21) and frequently during April. For example, increased water levels in wells South Oak, Wet Prairie, and Pin Oak from October 1997 through April 1998 coincided with relatively higher precipitation in those months (figs. 20–21). Groundwater levels then typically declined during the spring and summer months, except during months with high amounts of precipitation. Relatively wet years were distinguished in the water-level record by the absence of a decline during the growing-season months. This result indicates that the amount of precipitation in those periods exceeded water removed from the aquifer by evapotranspiration and by groundwater discharge from the aquifer to the Great Marsh. This result assumes that there is little or no vertical groundwater flow into or out of the surficial aquifer at Howe's Prairie.

Table 10. Summary statistics for groundwater levels (1985–2009) measured in five wells at Howe's Prairie near a natural part of the Great Marsh, northwestern Indiana. Well locations are shown in fig. 5.

[NGVD29, National Geodetic Vertical Datum of 1929]

Local well identifier	Years of data record	Number of water levels	Water-level altitudes, in feet above NGVD29				
			Minimum	25th percentile	Median	75th percentile	Maximum
North Oak	May 1985 to October 1993	64	597.41	598.1	598.99	600.54	602.26
	May to November 2009	16	600.95	601.76	602.09	602.55	602.84
Wet Prairie	April 1985 to November 2004	148	597.09	598.67	599.45	600.52	603.23
	May to November 2009	16	600.07	600.19	600.59	600.1	601.51
Pin Oak	May 1985 to November 1994	70	598.19	599.24	600.03	601.06	603.25
	May to November 2009	16	601.72	602.66	602.89	603.27	603.41
South Oak	April 1985 to November 1994	73	597.76	599.03	600.19	601.25	602.67
	May to November 2009	16	601.36	602.19	602.41	602.79	602.79
Well A	May 1989 to November 2004	129	596.47	597.51	598.54	599.54	602.77
	May to November 2009	16	600.91	601.67	601.97	602.47	602.59

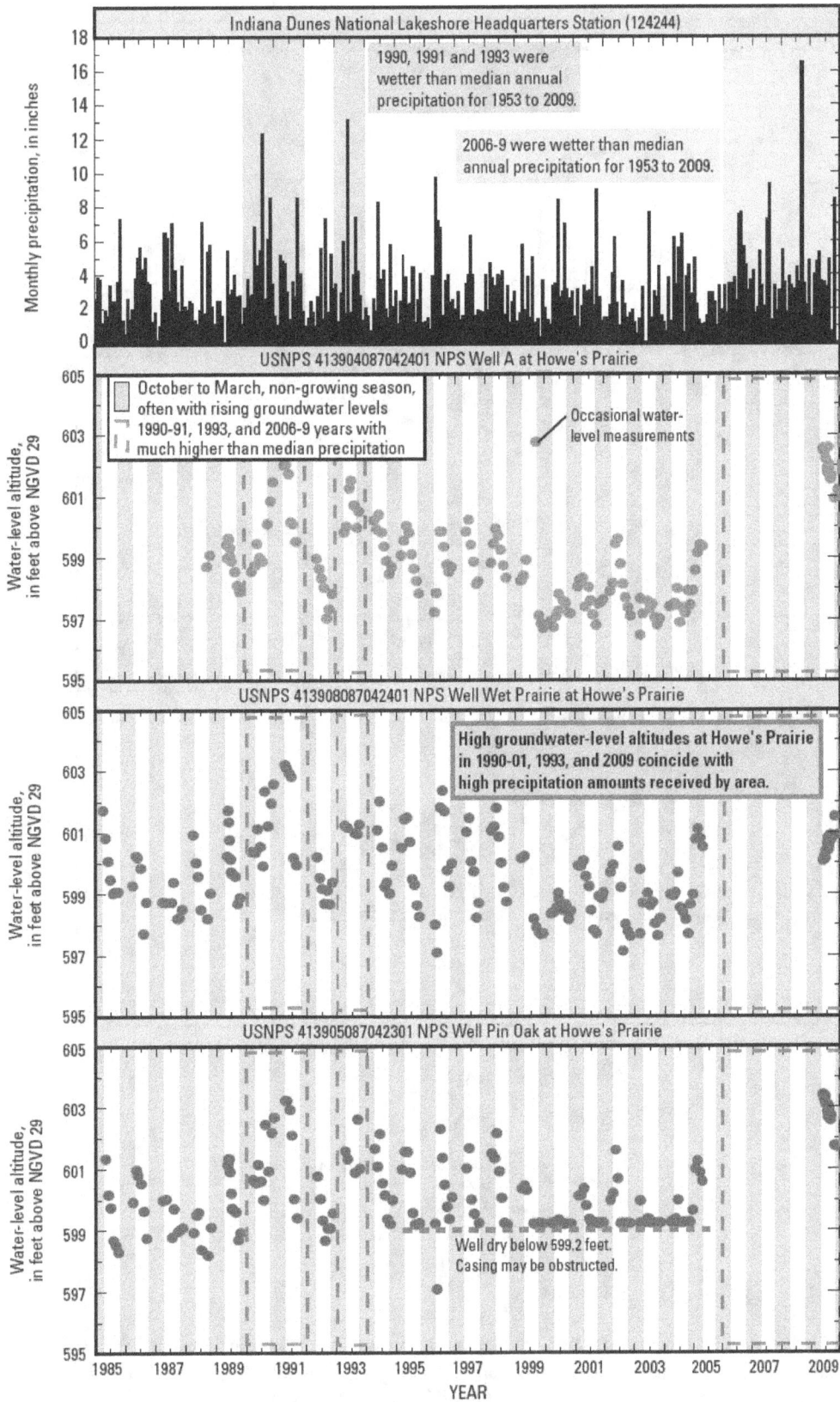

Figure 20. Precipitation and groundwater levels in observation wells A, Wet Prairie, and Pin Oak at Howe's Prairie near a natural part of the Great Marsh and precipitation amounts at Indiana Dunes National Lakeshore, northwestern Indiana, 1985–2009. Relatively higher groundwater levels in 1990, 1993, and 2009 correspond to periods of higher precipitation. Well locations are shown in fig. 5.

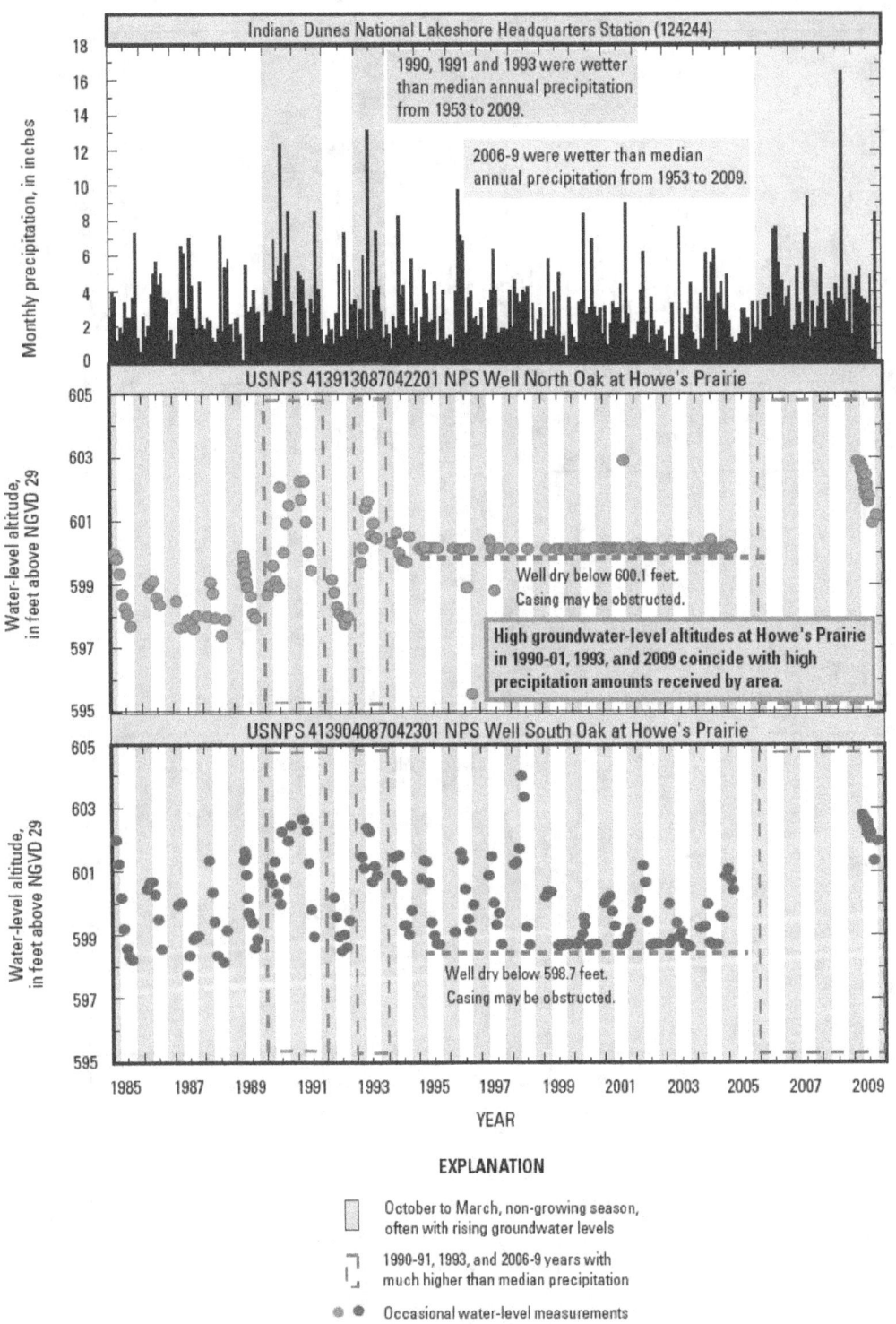

Figure 21. Precipitation and groundwater levels in observation wells North Oak and South Oak at Howe's Prairie near a natural part of the Great Marsh at Indiana Dunes National Lakeshore, northwestern Indiana, 1985–2009. Relatively higher groundwater levels in 1990, 1993, and 2009 correspond to periods of higher precipitation. Well locations are shown in fig. 5

Water levels in the Howe's Prairie wells in May to November 2009 were substantially higher than the range of water levels measured during 1985–2005. For example, the lowest (minimum) water level measured in all five wells in May to November 2009 was higher than the 75th percentile of water levels reported during 1985–2005 (table 10).

Water levels from three wells—North Oak during 1995–2004, South Oak during 1995–2004, and Pin Oak during 1994–2004—were excluded from the statistical comparison in table 10 because of obstructions in the well casings above the bottom of each well (figs. 20–21). The obstructions in these wells were likely from screen or casing failure and concurrent infiltration of aquifer sand into the screen; data from these periods are indicated as "casing may be obstructed" in figures 20–21. Nonetheless, water-level measurements that are above these apparent obstructions are considered to accurately reflect the natural water-table elevation because the aquifer sand that infiltrated into the screen and well casing below the obstructions is permeable to water.

Groundwater-Level Fluctuations After Wetland Restoration at Beverly Shores, 2007–9

Most of the relatively larger rises in water level of 0.5 ft or more in wells 511, 549, 551, 552, 553, 555, 556, 557, 560 and 562 were during or after rainfall and/or snowmelt. Groundwater-level fluctuations in these wells varied over a relatively narrow range of about 2 to 3 ft, with no net fluctuations greater than 4 ft in any well for the period of record (figs. 22–27). The same general pattern of water-level fluctuations was evident in most wells. Smaller, shorter-term rises in water level persisted over a few days to one or more weeks.

Water levels in several wells that are in the dune-beach complex (511, 549, 551, 553, 556, and 562) and adjacent to the Great Marsh (559B, 554) were within 0 to 6 ft of the land surface (figs. 22–27). Water levels at these sites indicate that basements in nearby residences could be within a depth that requires dewatering to maintain dry conditions. All but one of these wells (556) are in relatively low areas of the dune-beach complex: near a pond (562), the restored wetland (554), or interdunal wetlands (511, 549, 551, 553 and 559B) (fig. 17A). Water levels in wells 552 (fig. 19), 555 and 557 (figs. 22 and 23) were greater than 6 ft below land surface and indicate that basements in nearby residences were above a depth that requires dewatering to maintain dry conditions. These three wells are in relatively higher areas of the dune-beach complex as compared with wells near ponds or wetlands.

Seasonal and Event-Related Fluctuations

The pattern of water-level fluctuations in well 556 indicates the link between precipitation and recharge during the non-growing months and the sustained high groundwater levels through spring and early summer 2009. The water-level fluctuations at well 556 were typical of the seasonal variations in water levels in the dune-beach complex. Groundwater-level

fluctuations lasting days to weeks in the dune-beach complex were superimposed on a seasonal high water-table altitude. The water-table rise that created the seasonal high water-table altitude began with the recharge from snowmelt and rain in February 2009 and continued through July 2009. Water levels in well 556 were at a minimum near the end of the growing season and before post-growing-season rains, such as before the September 12–15, 2008, and in mid-September 2009 (fig. 22). Nearly immediate, sharp rises in groundwater levels were followed by relatively slow recessions after 1- to 2-day precipitation events of 1 in. or more in September, October, and December 2008 and in early March and late October 2009 (fig. 22). Sharp rises in water level in well 556 were followed by more rapid recessions during the 2009 growing season of April through September (fig. 22).

Water-level declines of about 0.1 to 0.4 ft/week in some dune-beach complex wells coincide with the growing season in late May through September. These declines relate to relatively higher amounts of evapotranspiration losses in the summer as compared with autumn. For example, water levels in well 562 (fig. 24) declined more slowly in October to December 2008 than in a period of comparable rainfall during June and July 2009. Evapotranspiration losses are less in this region during October to December as daily temperatures decline, the daylight period of each day decreases, and plant growth slows (Thornthwaite, 1948; Beaty, 1994). Daily maximum air temperatures were above 50°F for sustained periods from April through mid-October 2009 (fig. 24). Similar patterns related to evapotranspiration changes are seen in wells 511, 549, 551, 553, 555, 556, and 557 (figs. 22–23, 25–27).

Wetland and Drainage Effects on Water-Level Fluctuations

The lower hydraulic conductivity of organic sediments at the wetland margin may limit recharge and thereby explain the small groundwater level fluctuations in wells 554 and 559B after precipitation and after changes in nearby wetland surface-water levels. Water levels in some wells were generally stable, with fewer fluctuations than in other wells (well 559B, fig. 23; well 560, fig. 24; and well 554, fig. 27). These wells are either immediately north of the Great Marsh (wells 554 and 559B) or in a wetland area that was formerly part of the Great Marsh (well 560) (fig. 17A). The damping of water-level fluctuations in these wells probably is due to their proximity to the restored wetland and their being screened in or through organic sediments. Surface-water levels in the wetland areas immediately south of these wells and in Derby Ditch to the west of wells 559B and 551 represent the base levels for groundwater discharge. For example, water levels in wells 554 and 560 increase only during periods of high rainfall, such as those during mid-September 2008 and in late October 2009 (figs. 24 and 27). The response of water levels in these wells is slower and more subdued than water-level fluctuations at nearby surface-water sites SG1 and R504 (figs. 28 and 29, locations on fig. 17B).

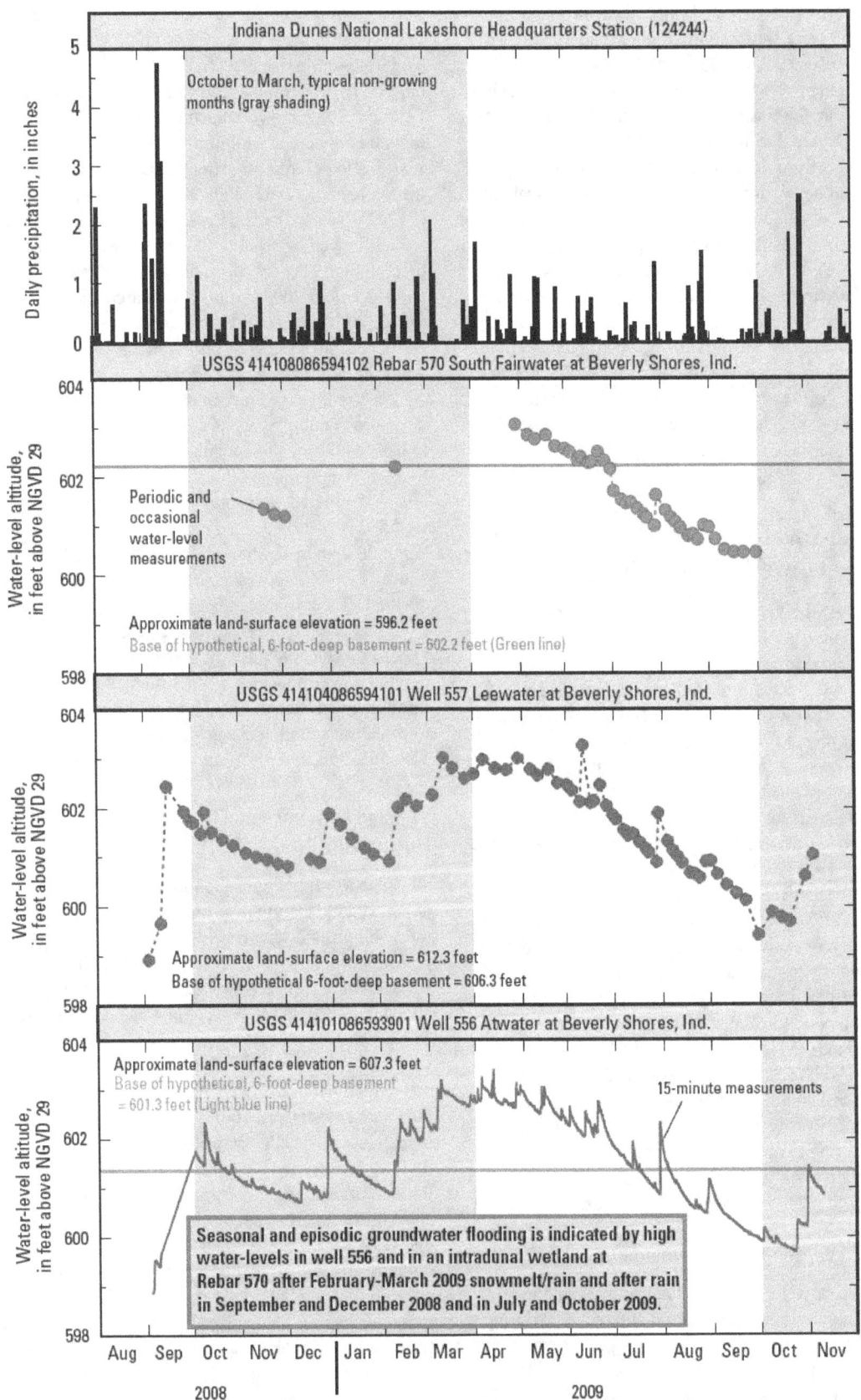

Note: Periodic water-level measurements not connected by dashed line if more than 10 days apart.

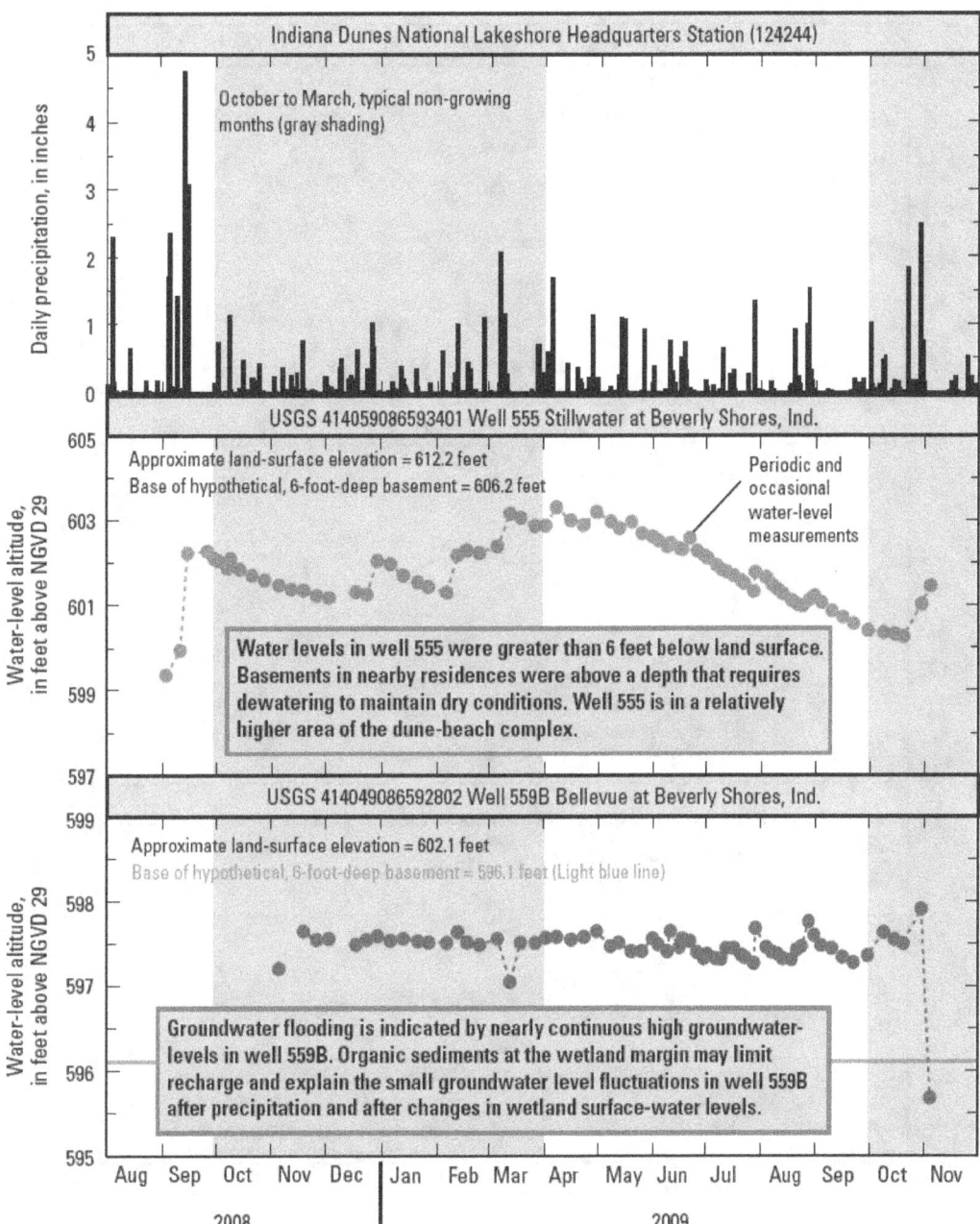

Note: Periodic water-level measurements not connected by dashed line if more than 10 days apart..

Figure 22. (left) Groundwater levels in wells 556 and 557 and in a wetland (Rebar 570) in the dune-beach complex at Beverly Shores along section *A–A´* relative to daily precipitation, northwestern Indiana, 2008–9. Larger water-level declines from May to September indicate the added effect of evapotranspiration on water levels. Well locations are shown in fig. 17*A*.

Figure 23. (above) Groundwater levels in wells 555 and 559B in the dune-beach complex at Beverly Shores along section *A–A´* relative to daily precipitation, northwestern Indiana, 2008–9. Larger water-level declines from May to September indicate the added effect of evapotranspiration on water levels. Well locations are shown in fig. 17*A*.

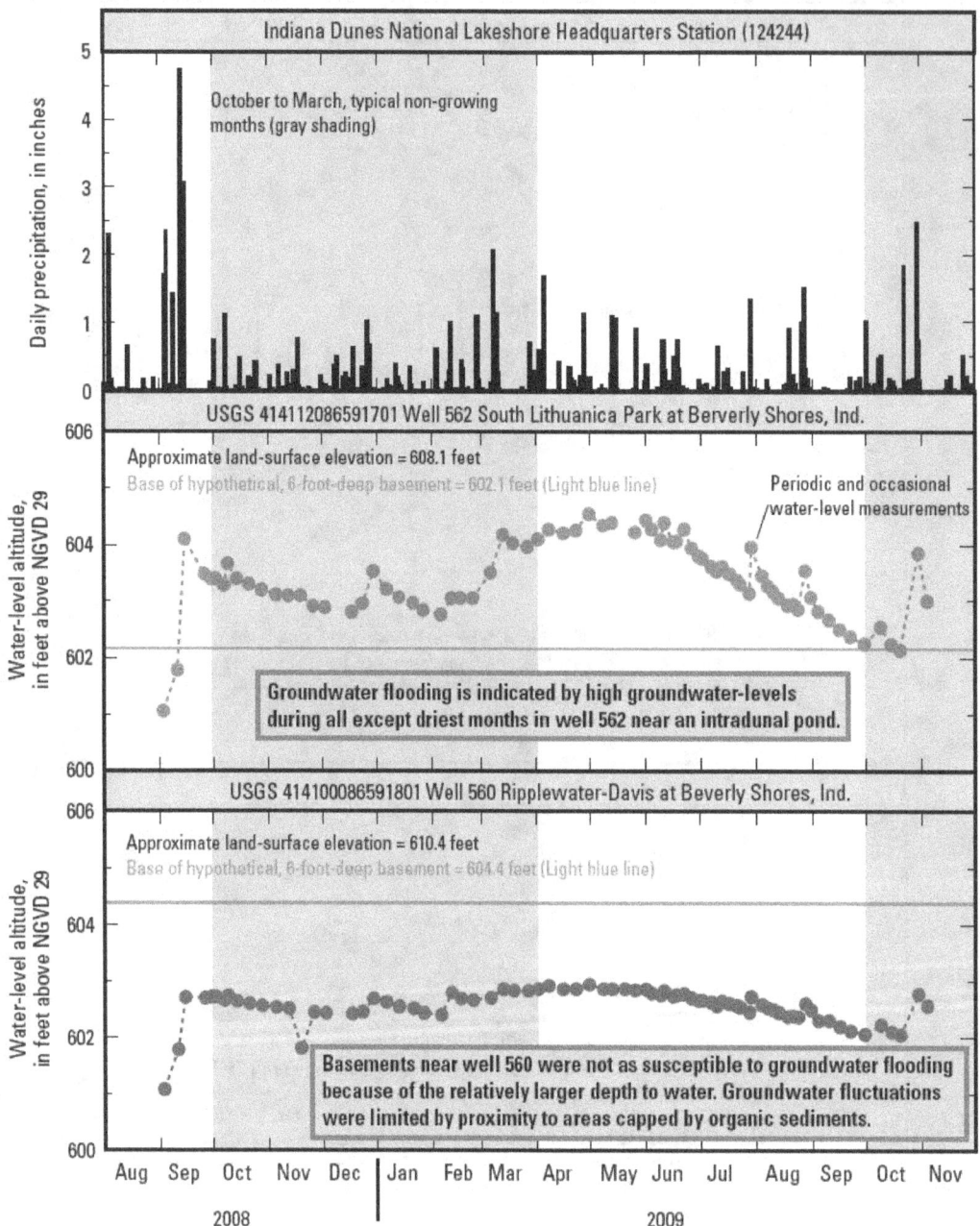

Figure 24. Groundwater levels in wells 560 and 562 in the dune-beach complex at Beverly Shores relative to daily precipitation, northwestern Indiana, 2008–9. Larger water-level declines from May to September indicate the added effect of evapotranspiration on water levels. Well locations are shown in fig. 17*A*.

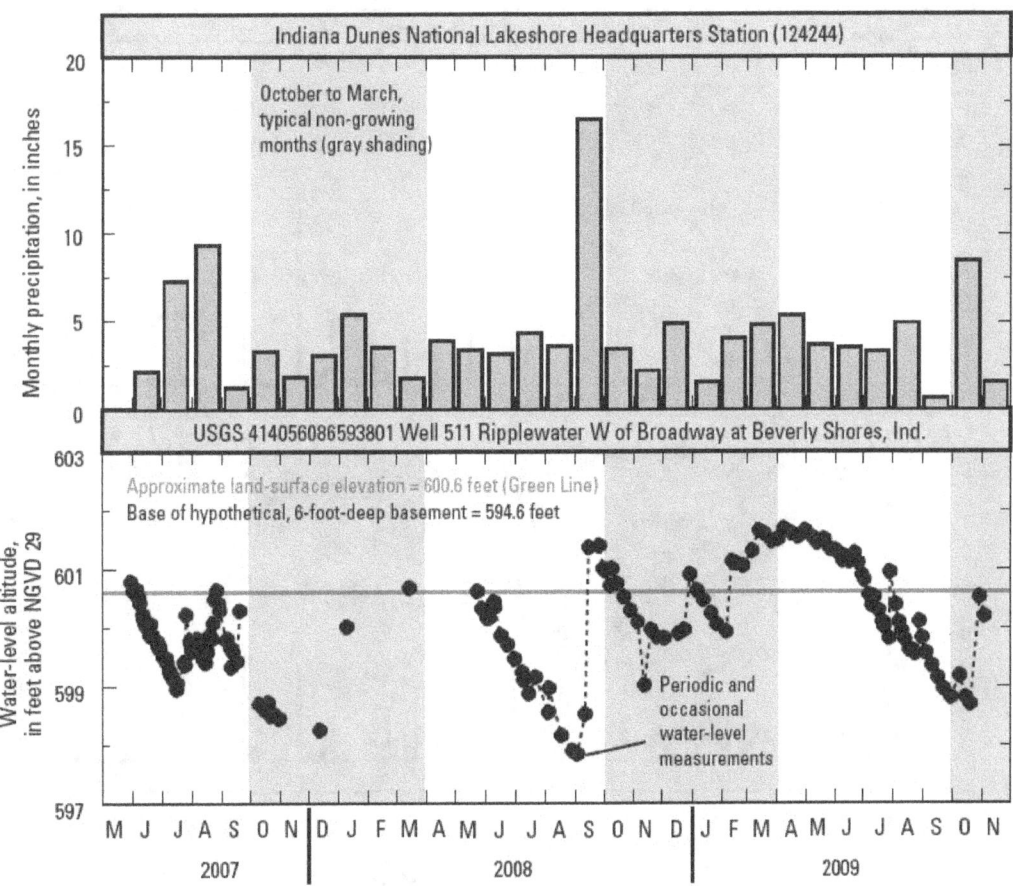

Figure 25.   Groundwater levels in well 511 in a wetland area within the dune-beach complex at Beverly Shores relative to monthly precipitation, northwestern Indiana, 2007–9. Larger water-level declines from May to September indicate the added effect of evapotranspiration on water levels. Well location is shown in fig. 17*A*.

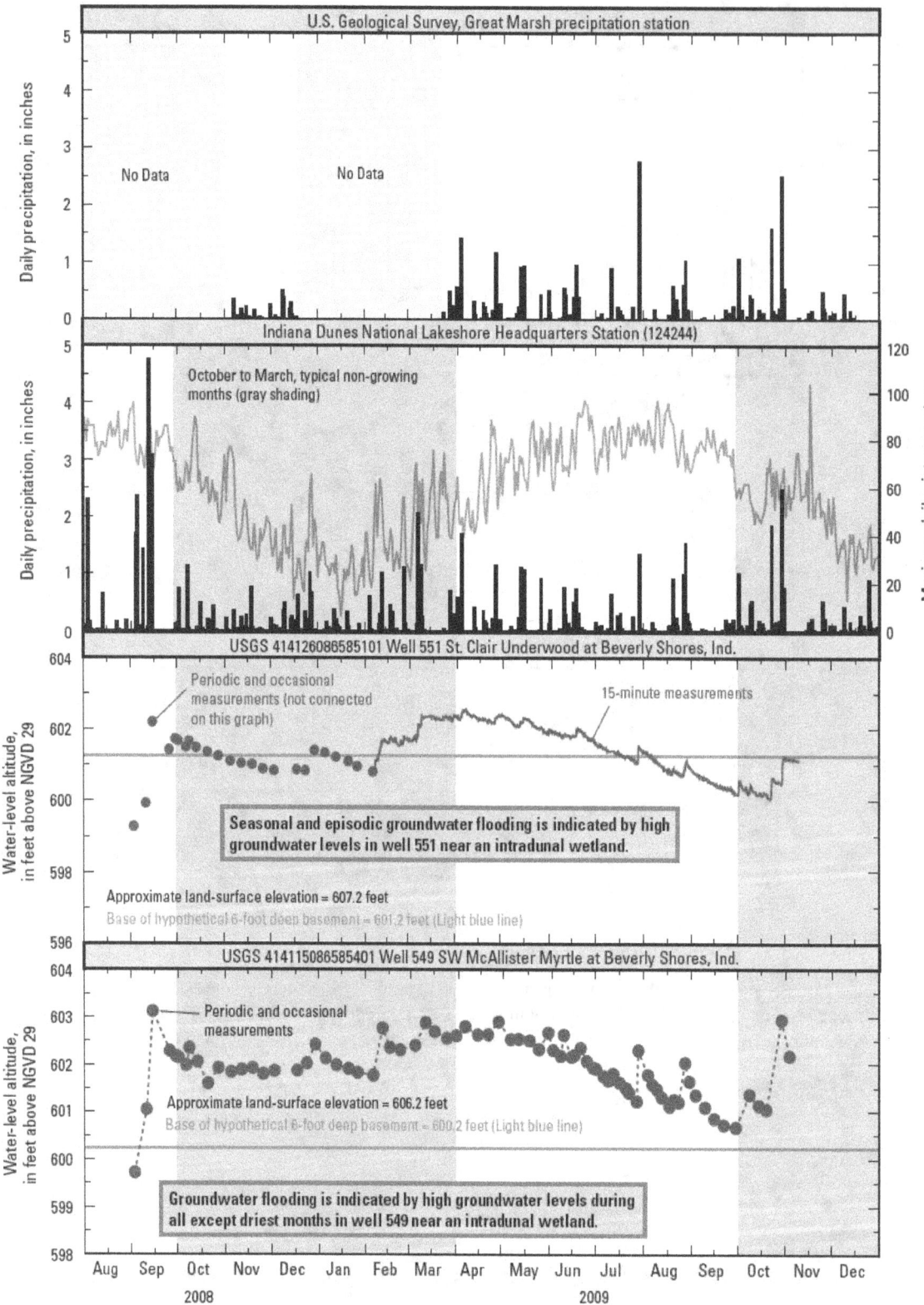

Figure 26.   Groundwater levels in wells 551 and 549 along section *B–B′* in the dune-beach complex at Beverly Shores relative to daily precipitation, northwestern Indiana, 2008–9. Larger water-level declines from May to September indicate the added effect of evapotranspiration on water levels. Well locations are shown in fig. 17*A*.

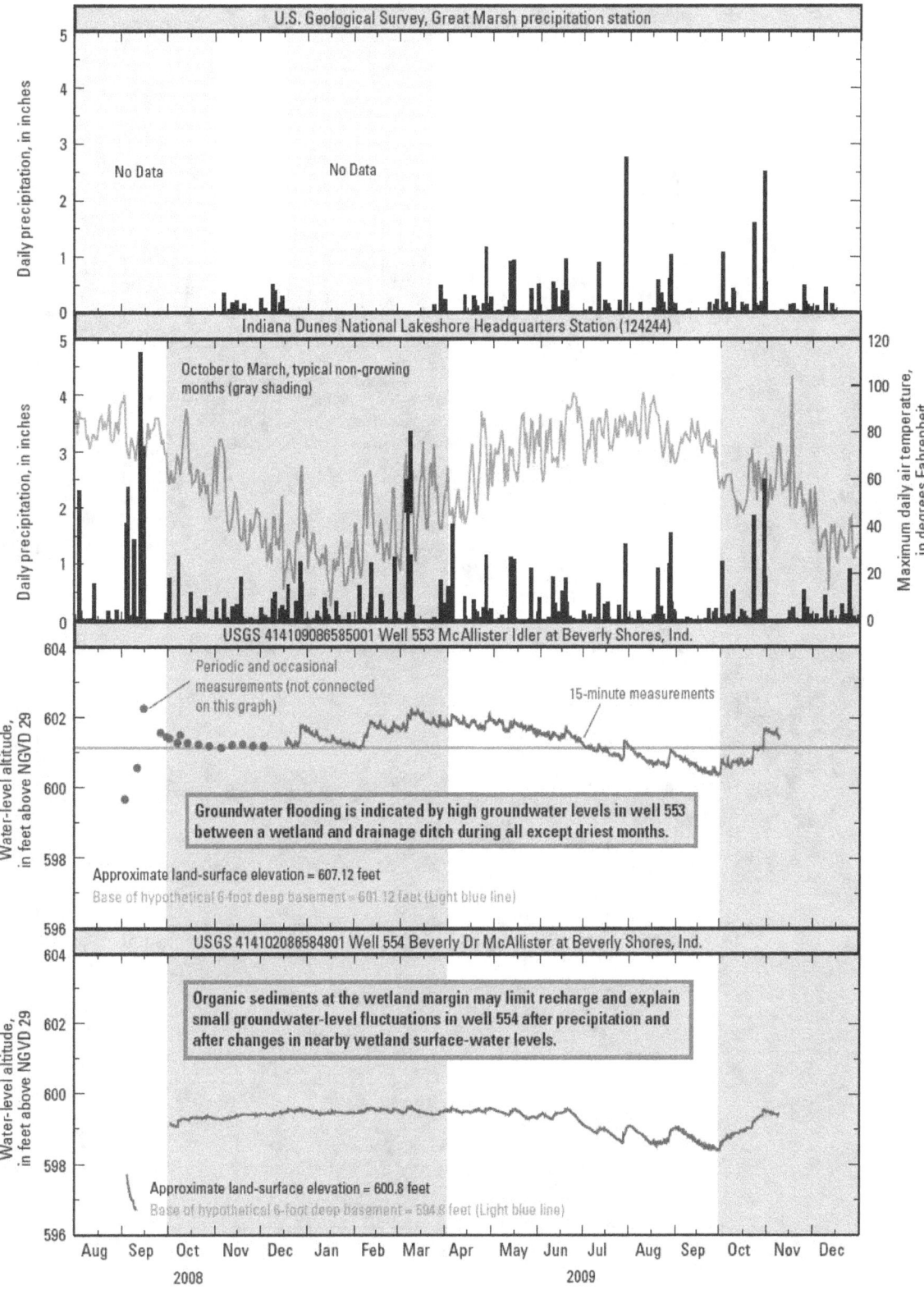

Note: Periodic water-level measurements not connected by dashed line if more than 10 days apart.

Figure 27.   Groundwater levels in wells 553 and 554 along section *B–B´* in the dune-beach complex at Beverly Shores relative to daily precipitation, northwestern Indiana, 2008–9. Larger water-level declines from May to September indicate the added effect of evapotranspiration on water levels. Well locations are shown in fig. 17*A*.

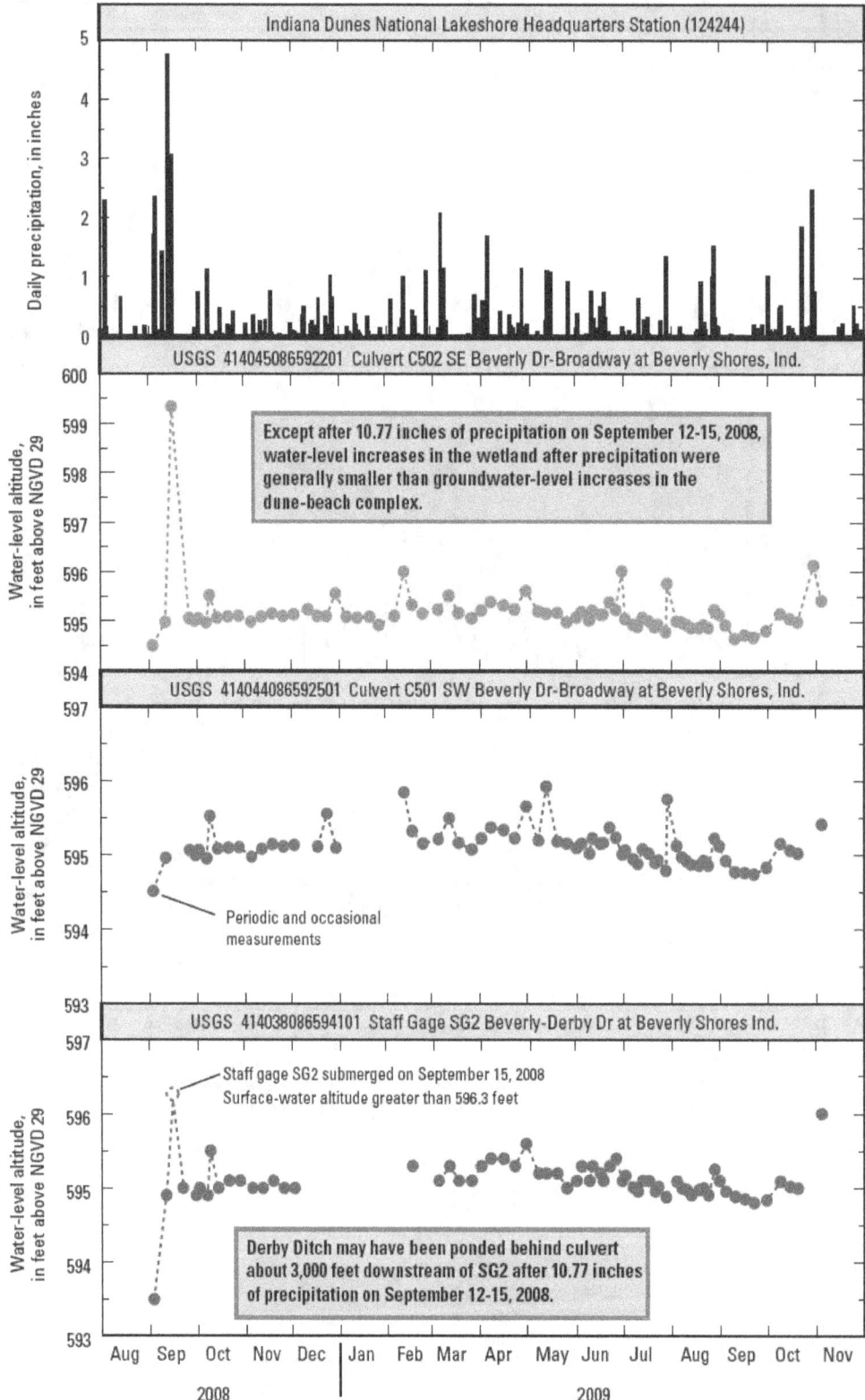

Figure 28. Surface-water-levels at sites C502, C501 and SG2 that are at and east of Broadway near Beverly Drive at Beverly Shores in a restored part of the Great Marsh, northwestern Indiana, 2008–9, relative to daily precipitation. Site locations are shown in fig. 17B.

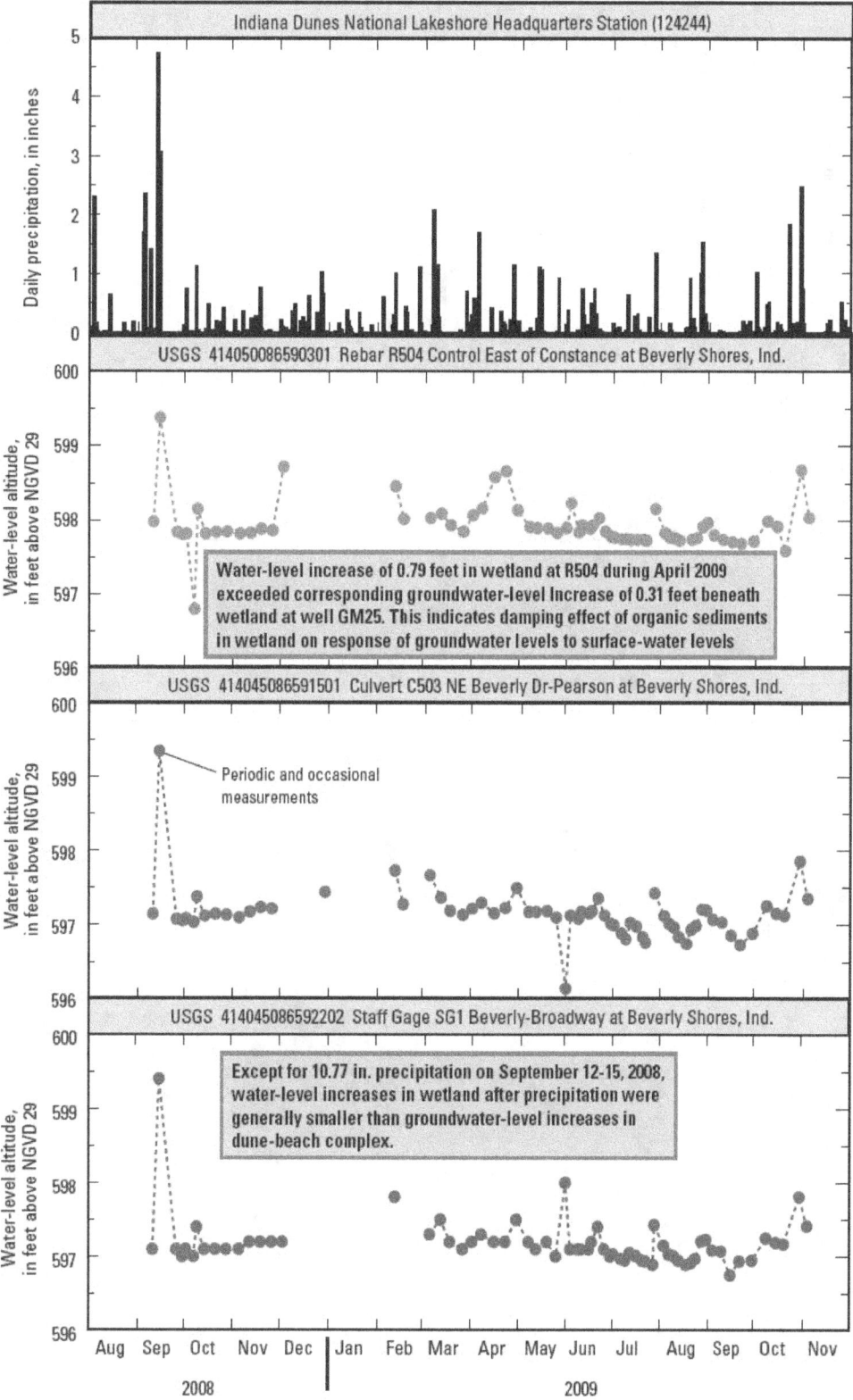

Figure 29.   Surface-water-levels at sites R504, C503 and SG1 between Broadway and well GM25 along Beverly Drive at Beverly Shores in a restored part of the Great Marsh, northwestern Indiana, 2008–9, relative to daily precipitation. Site locations are shown in fig. 17B.

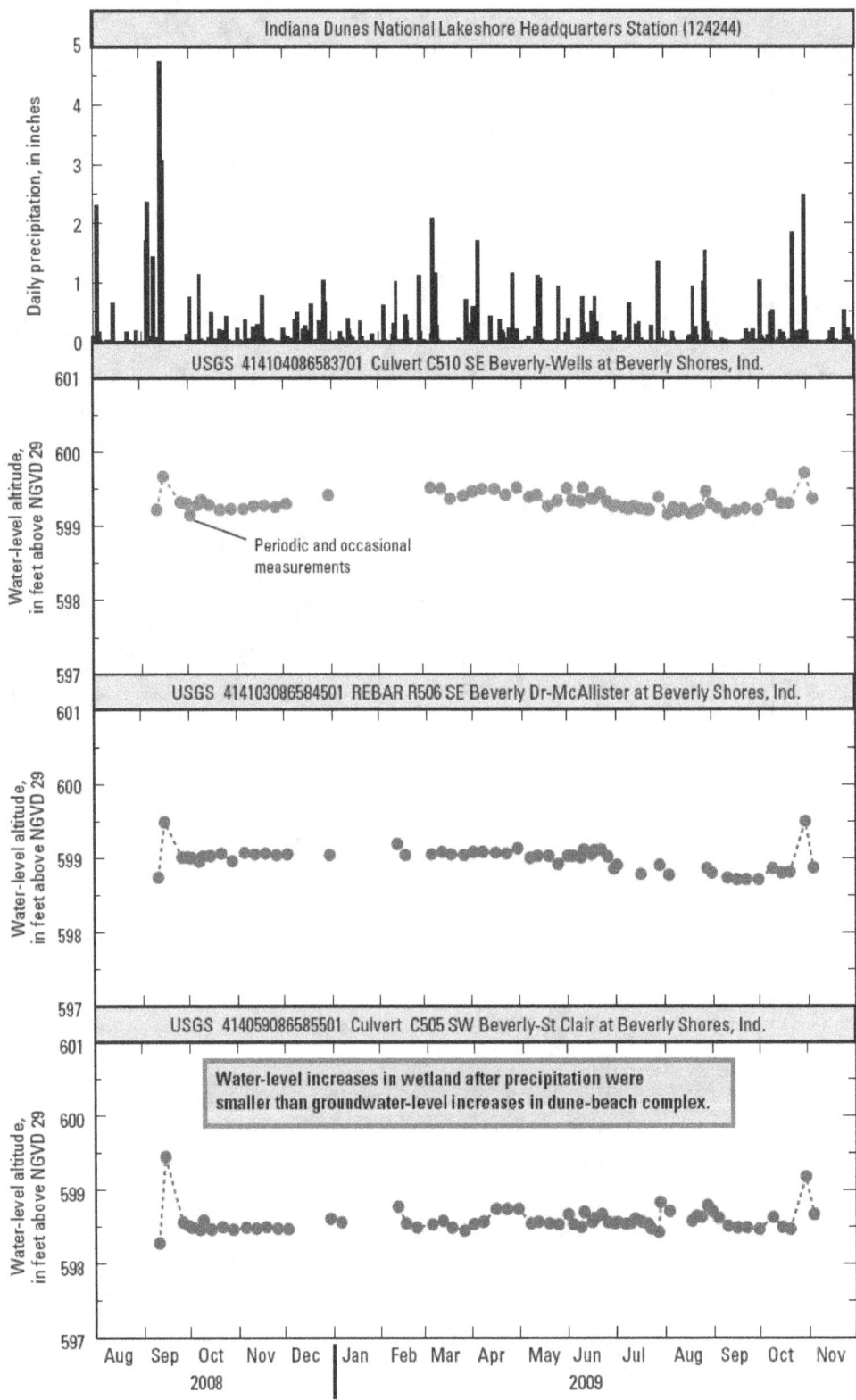

Figure 30.   Surface-water-levels at westernmost sites C510, R506, and C505 at Beverly Drive at Beverly Shores in a restored part of the Great Marsh, northwestern Indiana, 2008–9, relative to daily precipitation. Site locations are shown in fig. 17*B*.

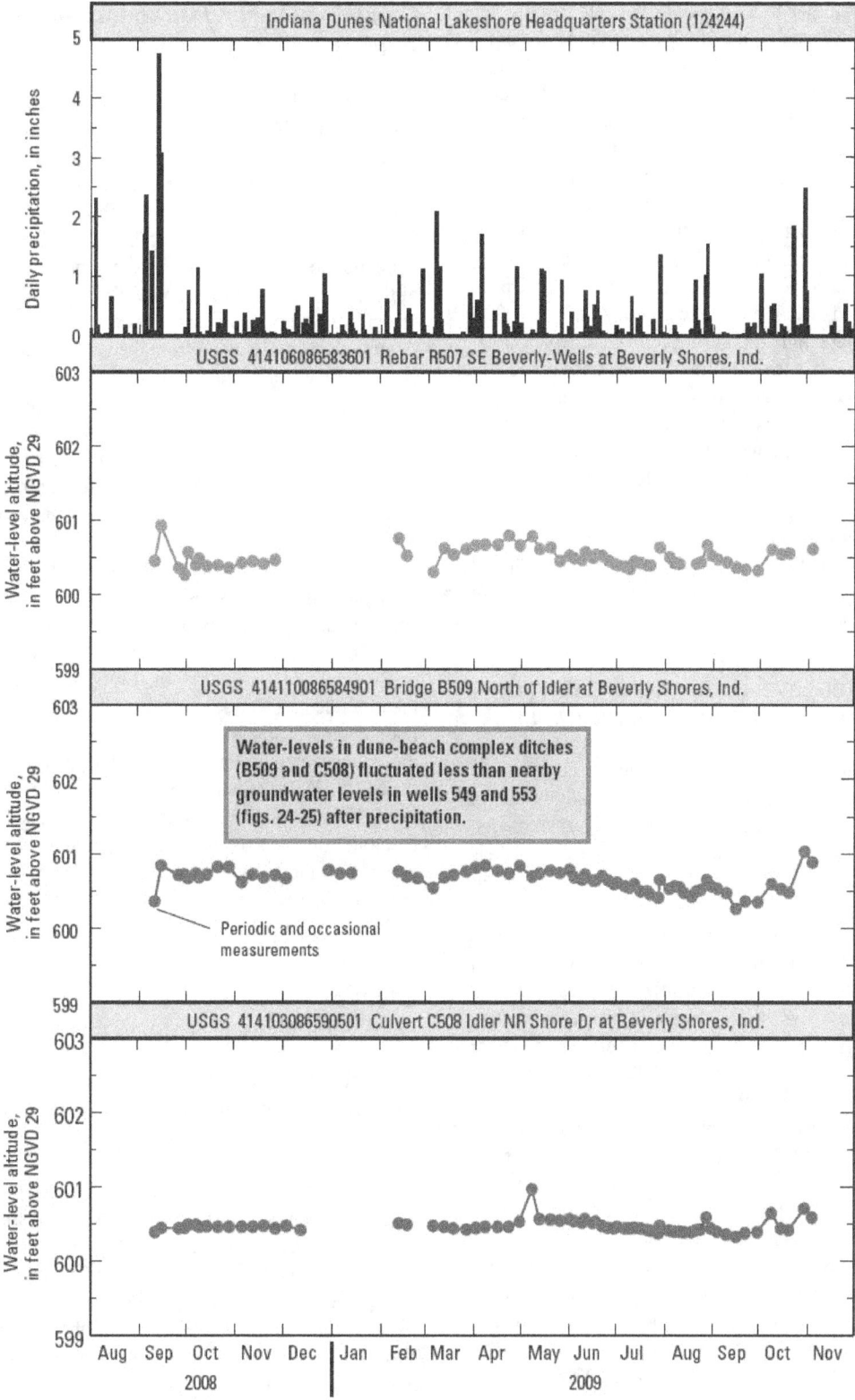

Figure 31. Surface-water-levels at sites R507, B509, and C508 in the dune-beach complex at Beverly Shores, northwestern Indiana, 2008–9, relative to daily precipitation. Site locations are shown in fig. 17*B*.

During the study period, flow was restored to a tile drain through Beverly Shores in July 2009; the drain flowed north in the subsurface along and near Broadway to about the area of site R570, then westward toward its discharge into Derby Ditch (fig. 17B). The effect of the flow restoration on rates of groundwater-level decline in July and August 2009 could not be distinguished in the 2008–9 water-level data for well 555 and surface-water site R570 (representing the local groundwater level in an intradunal wetland), the sites closest to the tile alignment. Water levels at these sites declined more rapidly, however, from the coincident effects of decreased precipitation and increased evapotranspiration starting in July 2009 and continuing until precipitation increased in October (figs. 22 and 23). Similar water-level declines at well 556 during June 2009 appeared mostly related to increased seasonal evapotranspiration (fig. 22). This pattern indicates that the effect of the tile drain on groundwater levels, if present, may not extend much beyond well 555 and site R570, about 250 and 100 ft from the tile alignment (fig. 17A). More water-level and evapotranspiration data are needed from these sites during winter months to identify whether the accelerated decline of groundwater-level continues during periods of less evapotranspiration and recharge.

Surface-Water-Level Fluctuations at Beverly Shores after Wetland Restoration, 2007–9

Surface-water-level fluctuations during this study varied over a narrower range than groundwater fluctuations, approximately from 1 to 1.5 ft, except for one set of measurements (figs. 28–31, site locations on fig 17B). Surface-water levels increased outside this range after the high rainfall in mid-September 2008, with water levels increasing the most at downstream sites and increasing the least at upstream sites. Surface-water levels at the upstream end of the restored wetland area in the Great Marsh (fig. 17B) at sites R507 and C510 rose only about 0.5 ft or less after the 10.77-in. rainfall on September 12–15, 2008 (figs. 30–31). Downstream in or near the restored wetland at sites R504, C505, and R506 (fig. 17B), surface-water elevations rose from about 1 to 1.5 ft (figs. 29–30). At the farthest downstream locations, sites C503 and C502 and staff gages SG1 and SG2, surface-water altitudes rose from about 2.5 ft to as much as 4.5 ft after the mid-September 2008 precipitation event (figs. 28–29). Site C501 could not be accessed during this event because of flooding. Surface-water levels at all sites except C508 and B509 returned to pre-event altitudes by the next measurement on September 26, 2008.

Water levels in dune-beach complex ditches (B509 and C508) fluctuated less after precipitation (fig. 31) than did nearby groundwater levels in wells 553 and 549. Surface-water levels along ditches in the dune-beach complex maintained a relatively consistent altitude during their 2008–9 period of measurement. The ditches in the dune-beach complex that are measured at sites C508 and B509 function as groundwater seeps from the surficial aquifer.

Groundwater-Flow Directions in the Surficial Aquifer

Perennial mounding of the water table in the surficial aquifer under the dune-beach complex indicates that the recharge that created the mound originates within the complex and not from flow from the adjacent hydrologic boundaries: the restored wetland, Lake Michigan, and Derby Ditch. Groundwater in the surficial aquifer flows from a water-table mound that is oriented approximately along the axis of the dune-beach complex beneath Beverly Shores toward groundwater discharges in topographically low areas (fig. 32). Groundwater flows away from the groundwater divide at the crest of the water-table mound through the surficial aquifer and discharges into areas with lower water levels—ditches in Beverly Shores, the restored wetlands to the south, Derby Ditch to the west, and Lake Michigan to the north. This and other broad, flat water-table mounds function as a groundwater divides under other topographically high parts of the dune-beach complex (Shedlock and others, 1994).

Water-table configurations mapped during this study all depict a mound of similar shape beneath the dune-beach complex during a variety of seasonally dry and wet hydrologic conditions in 2008 and 2009. After 10.77 in. of precipitation on September 12–15, 2008, the water-table altitude rose and the mounded area under the dune-beach complex expanded, but the general shape of the mound and groundwater-flow directions were similar to those before this precipitation event (figs. 32–33). Relatively greater increases in groundwater levels beneath the center of the dune-beach complex after precipitation and snowmelt as compared to smaller increases near the surface-water bodies at the complex's periphery indicate that recharge from precipitation and wastewater cause the mounding.

Effect of Recharge from Rain on Groundwater Levels and Flow Directions

Infiltrating precipitation causes most seasonal and episodic rises in groundwater levels beneath the dune-beach complex. As an example, groundwater levels in wells 551 and 553 along section B–B′ rose by 0.38 and 0.39 ft, respectively, in response to a precipitation event on October 1–2, 2009; 1.13 in. of rain was measured at the Great Marsh precipitation gage during that event (fig. 34). Slight rises in groundwater levels, about 0.04 ft at wells 551 and 553 and 0.02 ft at well 554, preceded the onset of precipitation on October 1, 2009 (fig. 34). The groundwater-level rose about 0.26 in. in well 556 starting at about 21:30 hours in response to the most intense, 1-in. part of the rainfall.

Rapid water-level rises in the restored wetland after precipitation do not likely have an effect on groundwater flooding elsewhere in the dune-beach complex. Time-delayed and smaller groundwater-level rises in wells 554 and 559B indicate a delaying effect on groundwater-level changes in and near the restored wetland from less conductive organic deposits in the subsurface near the marsh. Groundwater levels adjacent to the marsh at the two sites did not rise rapidly in response to short-term—less than 1 week—changes to surface-water levels in the restored part of the marsh. For example, a 2.74-in. rainfall on October 29–30, 2009, produced larger rises in groundwater levels at wells 551, 553, and 556 in the dune-beach complex than at well 554 near the restored wetland (fig. 35). Well 554 is about 100 ft from the restored wetland and wells 553, 551 and 556 are about 800 ft, 1,600 ft, and 2,000 ft, respectively from different parts of the restored wetland (fig. 17A). Groundwater levels at wells 551, 553, and 556 increased by 0.7 ft, 1.55 ft, and 1.23 ft, respectively, in response to the rainfall event. In contrast, groundwater levels at well 554 rose by only 0.18 ft (fig. 35). The water-level rise at well 554 was less than surface-water-level rises in response to the same event at sites R504 (1.09 ft), C505 (0.71 ft), and C510 (0.41 ft) in the restored wetland (figs. 29–30; site locations on fig. 17B). These changes in surface-water levels were measured near the end of rainfall on October 30 between about 15:00 and 16:30 hours. A similar lag in water-level rise near the wetland was observed in weekly measurements of water levels in well 559A after the mid-September 2008 precipitation (fig. 32–33). Well 559A was replaced with the slightly deeper well 559B in September 2008.

After about 10.77 in. of rain fell at the INDU station from September 12–15, 2008, water levels at all wells rose; however, water levels in wells at relatively lower areas inside the dune-beach complex rose the most; by about 2.3 ft at well 562 and 2.1 ft at well 549 (figs. 24 and 26, 32–33). These relatively larger water-level rises were next to an interdunal pond at well 562 and in an interdunal wetland area at well 549 where the depth to the water table was lowest. This sequence is similar to that described by Winter (1999, p. 32), in that recharge, when evenly applied, first reaches the water table where the unsaturated zone is thin relative to adjacent areas. Water table increases from recharge then progress laterally over time to areas with thicker unsaturated zones. The water-table mound altitude under the dune-beach complex later decreases as groundwater flows from the mound toward adjacent discharge areas (Winter, 1999, p. 33). A partial exception to that analogy may have arose if surface-water drainage from the restored wetland backed up behind the culvert from Derby Ditch to Lake Michigan. That condition, if present, would have limited changes in surface-water levels along Derby Ditch west of section A–A´ (fig. 31, section location on fig. 4).

Many of the sharp rises in the water table observed in spring months or after periods with frequent rain could have been affected by changes in the partial saturation of unsaturated zone porosity. For example, a hypothetical infiltration of the 10.77-in. rainfall (September 2008) in an aquifer with an effective porosity of 0.3, after accounting for consumptive losses by absorption of some water into plant matter (say 10 percent, leaving 9.7 in. for infiltration), could lead to a hypothetical rise in the water table of about 2.7 ft. If subsurface conditions were drier, as would be expected during the low water table conditions before the September 12-14, 2008

rainfall, a slightly larger effective porosity of 0.35 corresponding to a drier unsaturated zone would produce an equivalent water table rise of about 2.3 ft. In comparison, the largest water table rises observed after the September 12-14 rain were about 2.3 ft at well 562 and 2.1 ft at well 549 (figs. 24, 26, 32–33). Relatively wetter antecedent conditions such as after frequent rain and infiltration could decrease the effective porosity and increase the amount of water table rise per inch of recharge. The same hypothetical infiltration of the 10.77-in. rainfall (September 2008) in an aquifer with a smaller effective porosity of 0.2, after accounting for the same consumptive losses as above, could lead to a much larger rise in the water table of about 4 ft.

Recharge from 2.75 in. of precipitation on July 28, 2009, caused changes to the water-table configuration similar to those from previous, relatively high rainfalls in the area. The wells with 1 ft or more of rise in groundwater level in response to the precipitation were wells 556 (1.22 ft), 511 (1.12 ft), 549 (1.08 ft), and 557 (1 ft) (figs. 36–37). The area of the rise in water-table altitude is largely indicated by the increase in area included in the 602-ft contour (figs. 36–37). Three of these four wells were along section A–A´; this increase in area indicates a greater relative rise in the water table from this storm along the western end of the study area.

Effect of Recharge from Snowmelt and Rain on Groundwater Levels and Flow Directions

The seasonal rise in dune-beach complex groundwater levels is affected by vertical recharge from snowmelt and rain and substantially not by changes in groundwater levels near wetland. Recharge from infiltrating snowmelt and precipitation during February and March 2009 caused the water table to rise beneath the dune-beach complex between the Great Marsh and Lake Michigan (figs. 38–41). That recharge did not cause an appreciable water-table rise at wells in the Great Marsh. Water-table configurations before and after a major snowmelt and rain in February 2009 changed in a similar manner to those before and after the September 2008 precipitation event. The February 6, 2009, water table before the snowmelt and precipitation event (fig. 38) had a broad mound centered on wells 560, 562, 549, and 552 that was similar to the pre-rainfall configuration on September 11, 2008 (fig. 30). During February 7–14, 2009, daily high temperatures rose substantially above freezing (fig. 39). The higher temperatures and snowmelt on February 7–8, and rainfall on February 10–12 resulted in infiltration to the surficial aquifer that was observed as increased water levels in most wells. An example of this increase is indicated by about a 0.4- to 0.8-ft increase in water levels at well 553 from February 6 through March 1, 2009 (fig. 39). During the same period, water levels in well 554 adjacent to the Great Marsh lowered slightly, from 597.41 to 597.34 ft (fig. 39). A March 6, 2009, water-table map from after the snowmelt and precipitation (fig. 40) indicated increased (higher) water levels and a larger, broader water-table mound under the dune-beach complex.

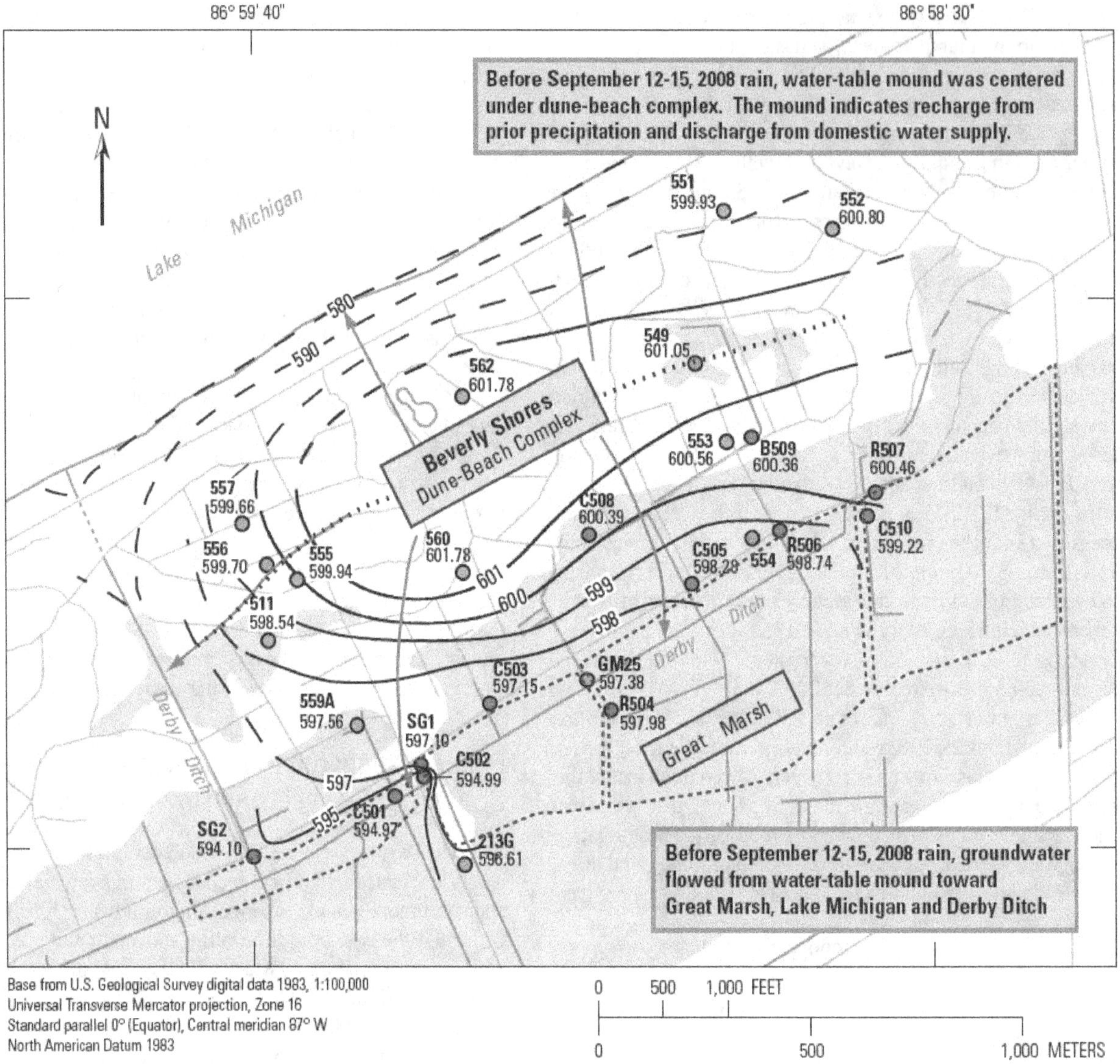

Figure 32.   Water-table contours and directions of groundwater flow in the surficial aquifer, September 11, 2008, during relatively dry conditions before the Hurricane Ike-Tropical Storm Lowell precipitation event of September 12–15, 2008, northwestern Indiana.

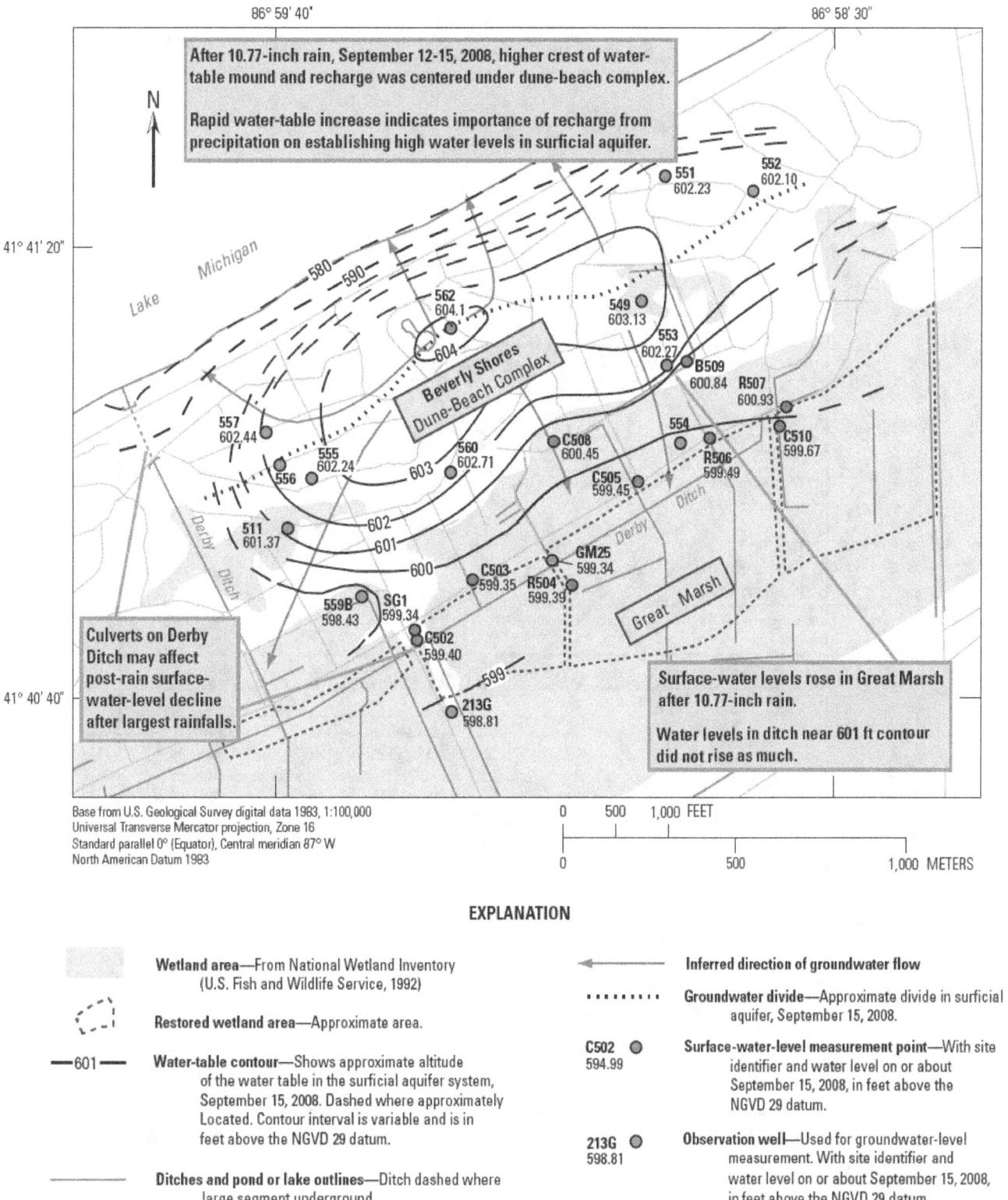

86° 59' 40"

86° 58' 30"

N

After 10.77-inch rain, September 12-15, 2008, higher crest of water-table mound and recharge was centered under dune-beach complex.

Rapid water-table increase indicates importance of recharge from precipitation on establishing high water levels in surficial aquifer.

41° 41' 20"

Lake Michigan

580
590

551 602.23

552 602.10

562 604.1

Beverly Shores Dune-Beach Complex

604

549 603.13

553 602.27

B509 600.84

R507 600.93

557 602.44

555 602.24

556

560 602.71

603

554

C508 600.45

R506 599.49

C510 599.67

Ditch

511 601.37

602

601

C505 599.45

Derby Ditch

600

C503 599.35

R504 599.39

GM25 599.34

Great Marsh

Culverts on Derby Ditch may affect post-rain surface-water-level decline after largest rainfalls.

559B 598.43

SG1 599.34

C502 599.40

599

41° 40' 40"

213G 598.81

Surface-water levels rose in Great Marsh after 10.77-inch rain.

Water levels in ditch near 601 ft contour did not rise as much.

Derby Ditch

Base from U.S. Geological Survey digital data 1983, 1:100,000
Universal Transverse Mercator projection, Zone 16
Standard parallel 0° (Equator), Central meridian 87° W
North American Datum 1983

0   500   1,000 FEET

0   500   1,000 METERS

EXPLANATION

Wetland area—From National Wetland Inventory (U.S. Fish and Wildlife Service, 1992)

Restored wetland area—Approximate area.

—601— Water-table contour—Shows approximate altitude of the water table in the surficial aquifer system, September 15, 2008. Dashed where approximately Located. Contour interval is variable and is in feet above the NGVD 29 datum.

Ditches and pond or lake outlines—Ditch dashed where large segment underground.

Inferred direction of groundwater flow

Groundwater divide—Approximate divide in surficial aquifer, September 15, 2008.

C502 599.34 Surface-water-level measurement point—With site identifier and water level on or about September 15, 2008, in feet above the NGVD 29 datum.

213G 598.81 Observation well—Used for groundwater-level measurement. With site identifier and water level on or about September 15, 2008, in feet above the NGVD 29 datum.

Figure 33. Water-table contours and directions of groundwater flow in the surficial aquifer, September 15, 2008, during relatively wet conditions 1 day after the Hurricane Ike-Tropical Storm Lowell precipitation event of September 12–15, 2008, northwestern Indiana.

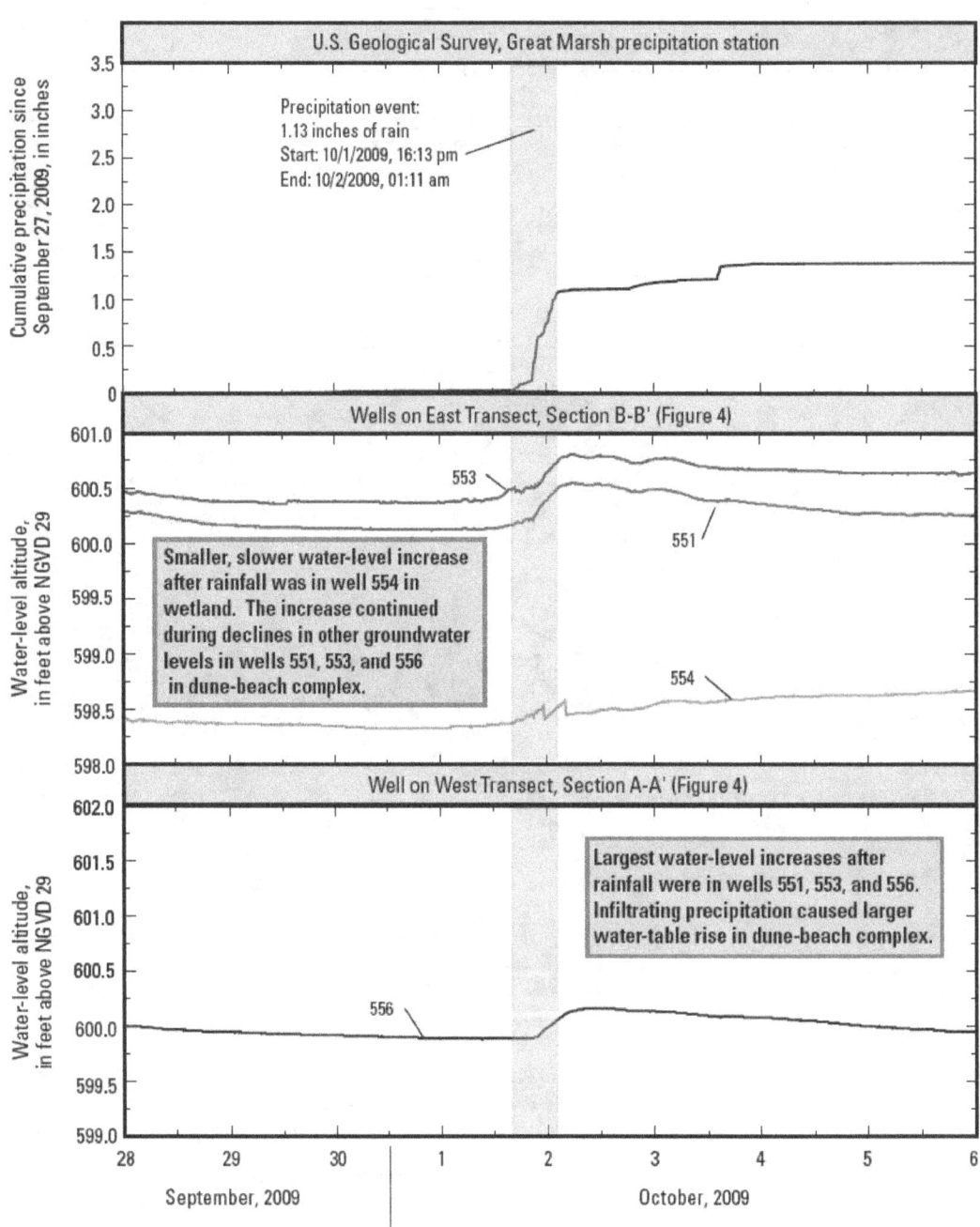

Figure 34.   Rise in groundwater levels related to precipitation on October 1–2, 2009, at wells 551, 553, and 556 in the dune-beach complex and well 554 near a restored part of the Great Marsh, northwestern Indiana. Well locations are shown in fig. 17A.

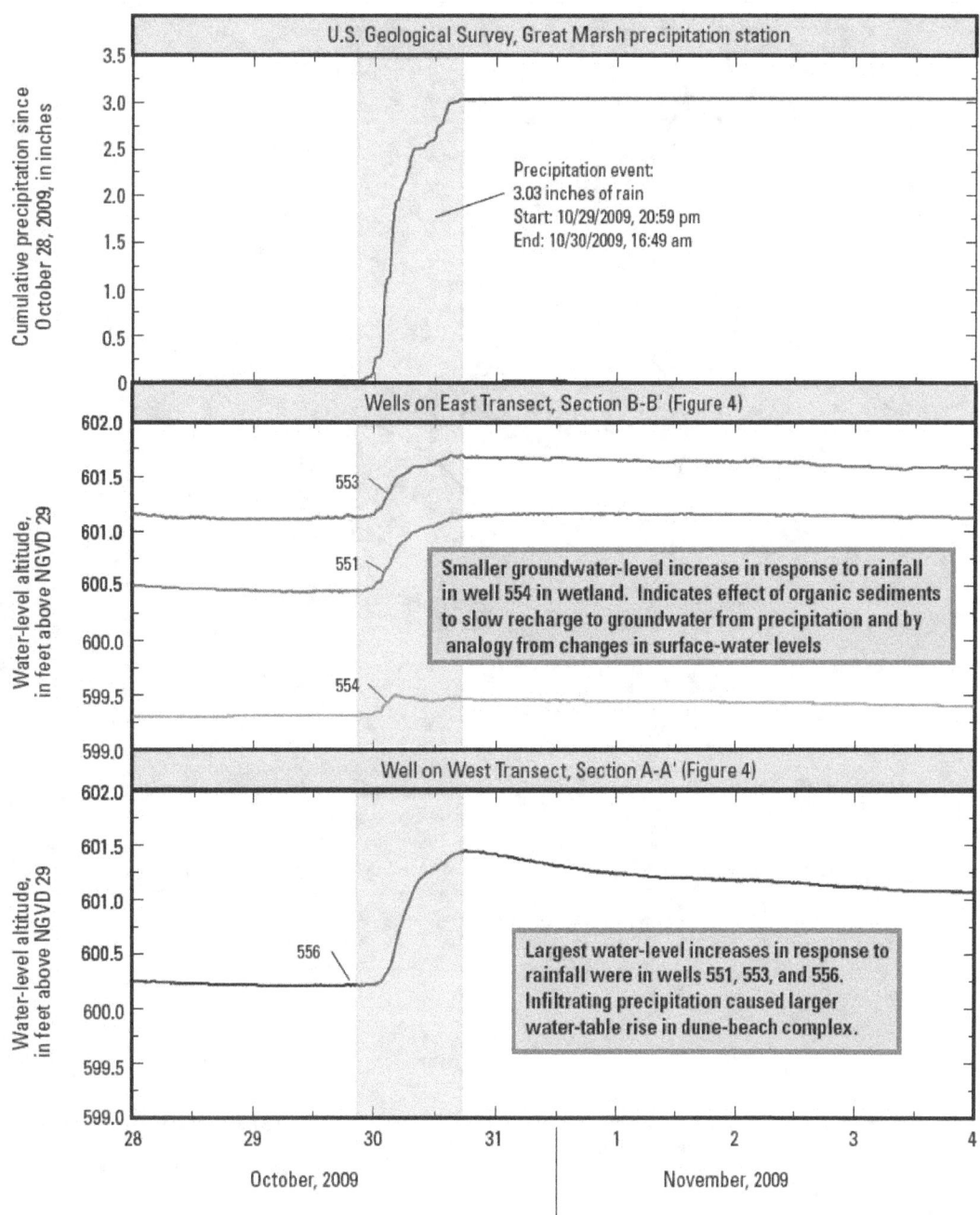

Figure 35. Rise in groundwater levels related to precipitation on October 29–30, 2009, at wells 551, 553, and 556 in the dune-beach complex and well 554 near a restored part of the Great Marsh, northwestern Indiana.

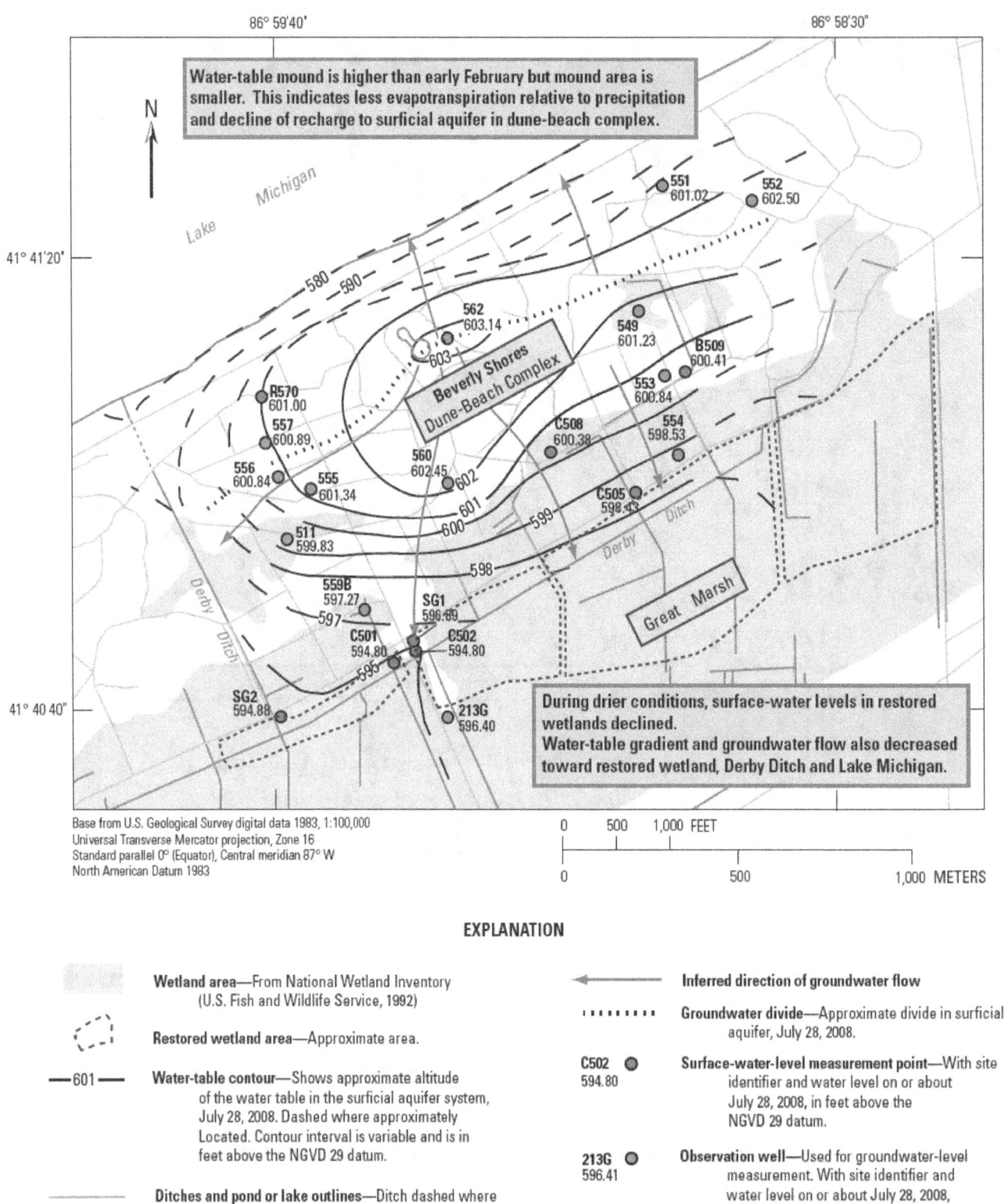

Figure 36.   Water-table contours and directions of groundwater flow in the surficial aquifer, July 28, 2009, before 2.75 inches of precipitation fell at the Great Marsh precipitation station, northwestern Indiana.

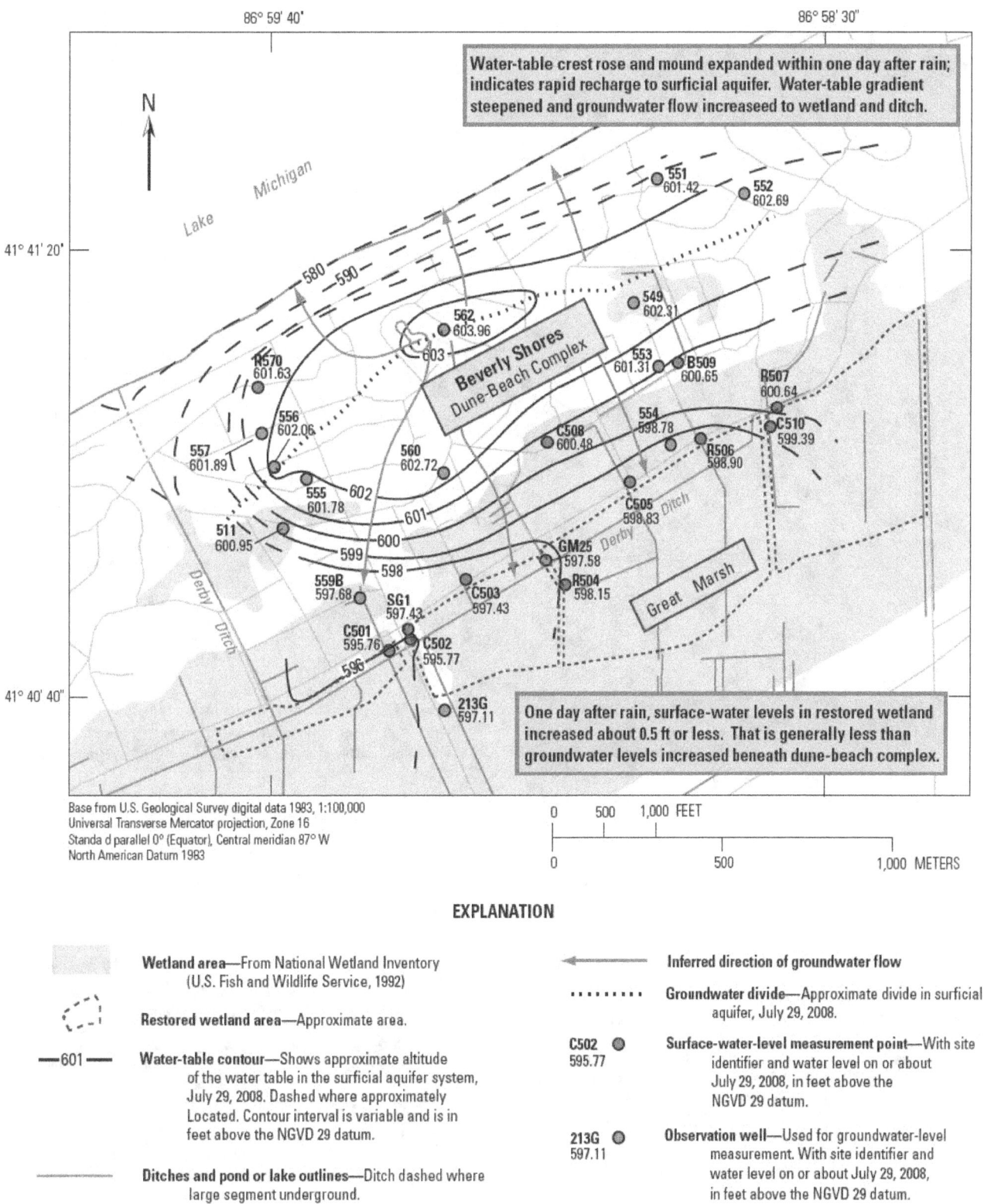

Figure 37. Water-table contours and directions of groundwater flow in the surficial aquifer, July 29, 2009, after 2.75 inches of precipitation fell at the Great Marsh precipitation station on July 28, 2009, northwestern Indiana.

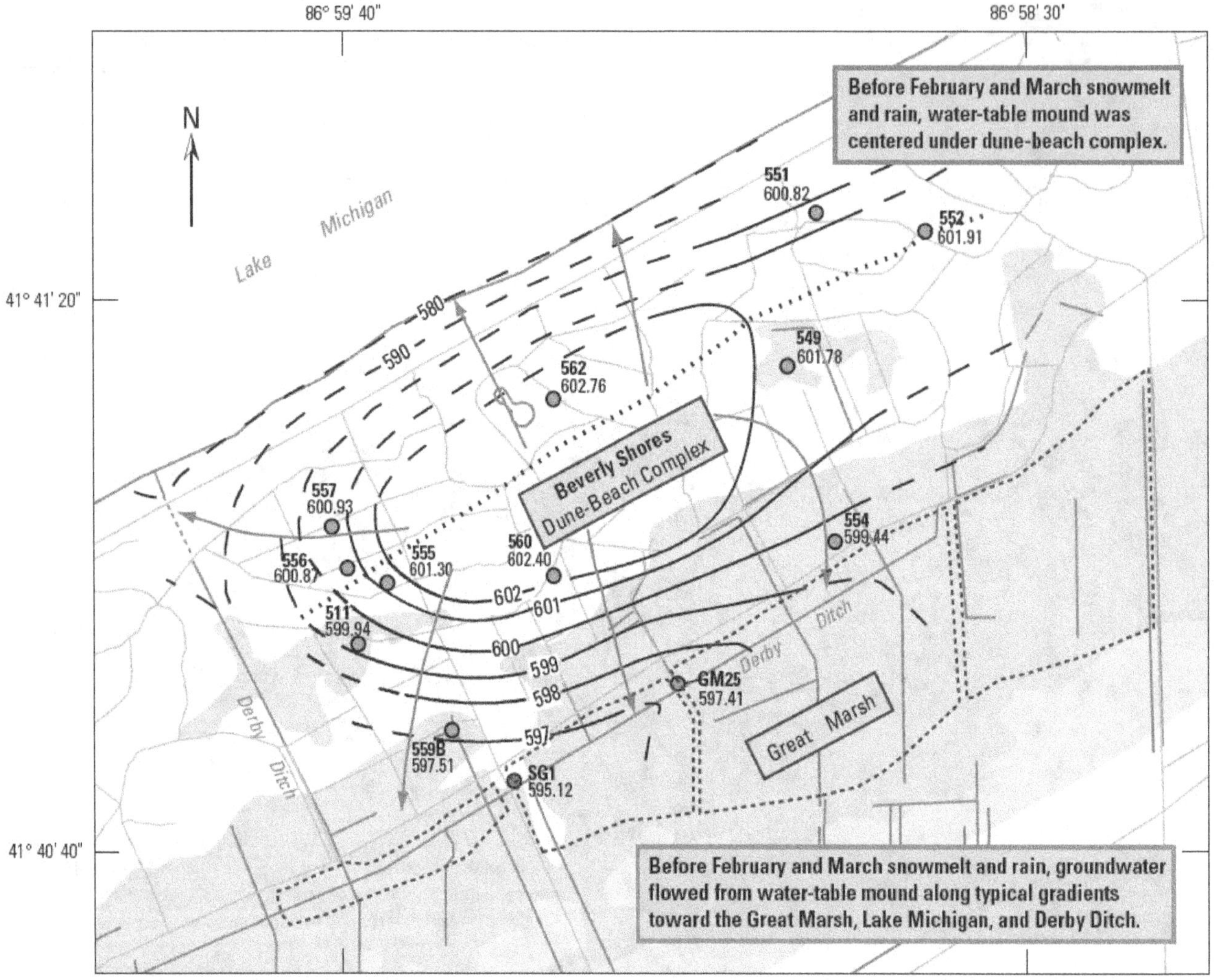

Figure 38.   Water-table contours and directions of groundwater flow in the surficial aquifer, February 6, 2009, before snowmelt and precipitation events, northwestern Indiana.

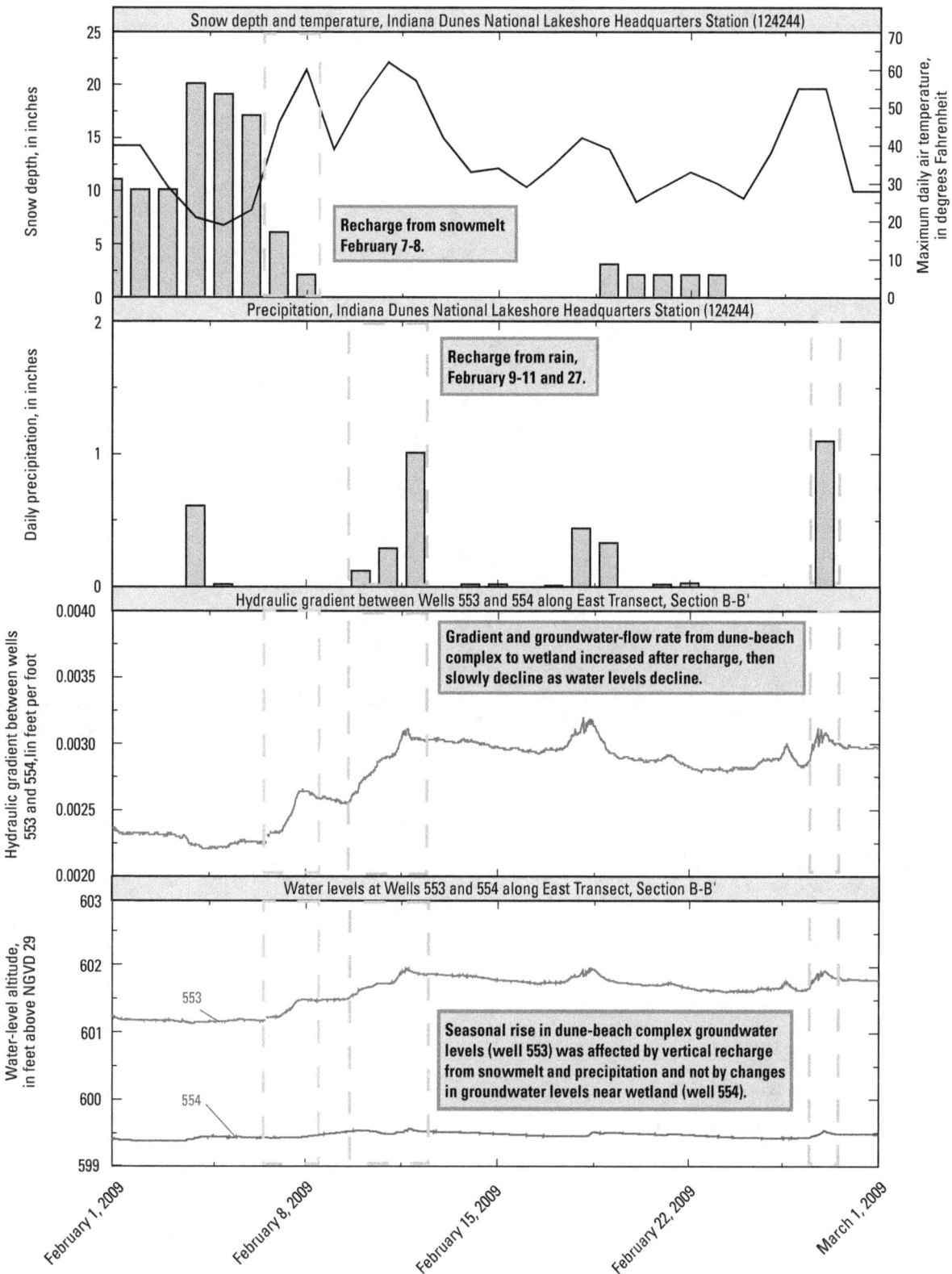

Figure 39. Relation of snow thickness, rainfall, and daily mean temperature at the Indiana Dunes National Lakeshore weather station to changes in water levels in well 553 on a dune ridge and well 554 near a restored area of the Great Marsh, northwestern Indiana, during February 2009.

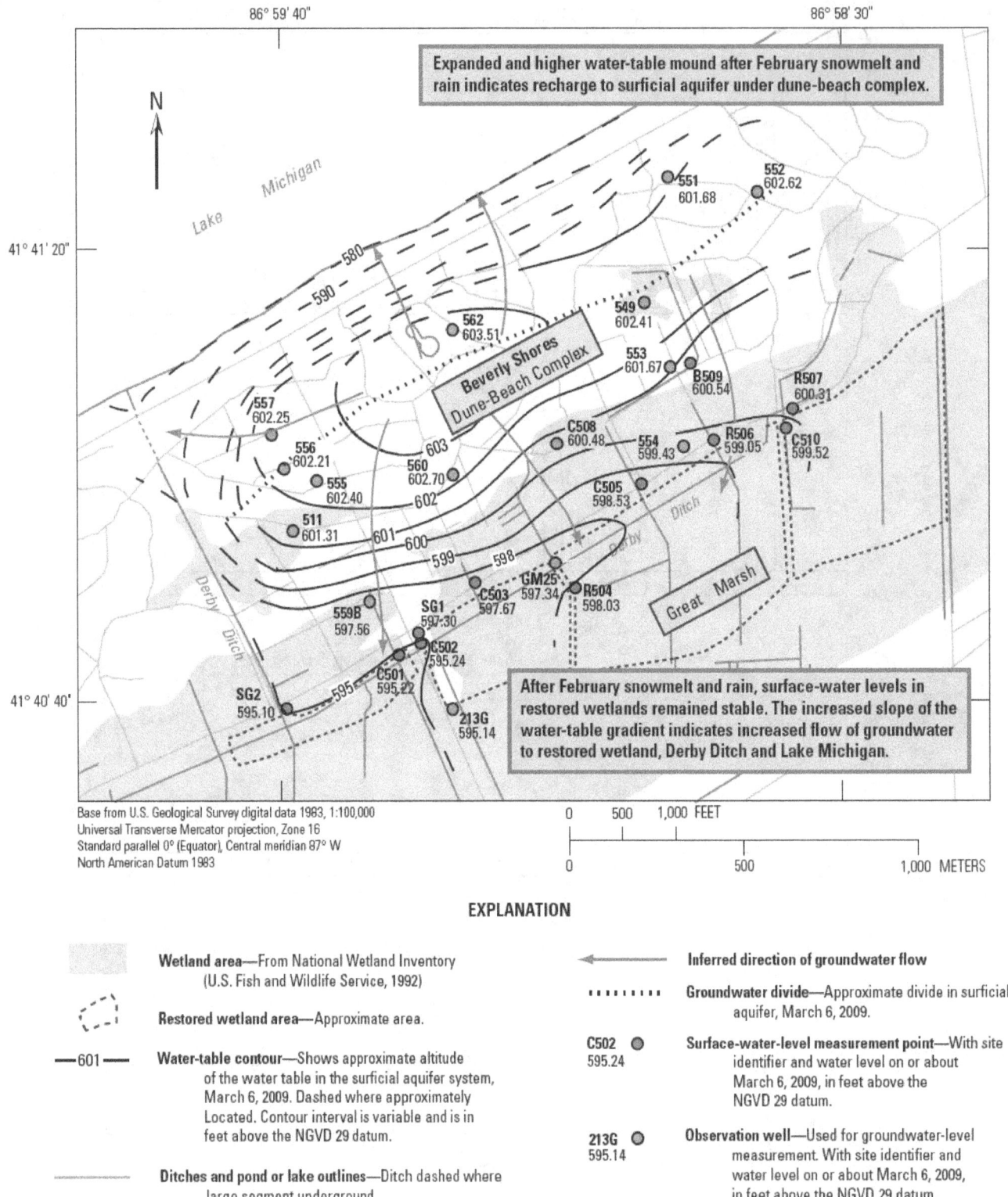

Figure 40.   Water-table contours and directions of groundwater flow in the surficial aquifer, March 6, 2009, after a major snowmelt and corresponding rise in the water table beneath the dune-beach complex area between the Great Marsh and Lake Michigan, northwestern Indiana.

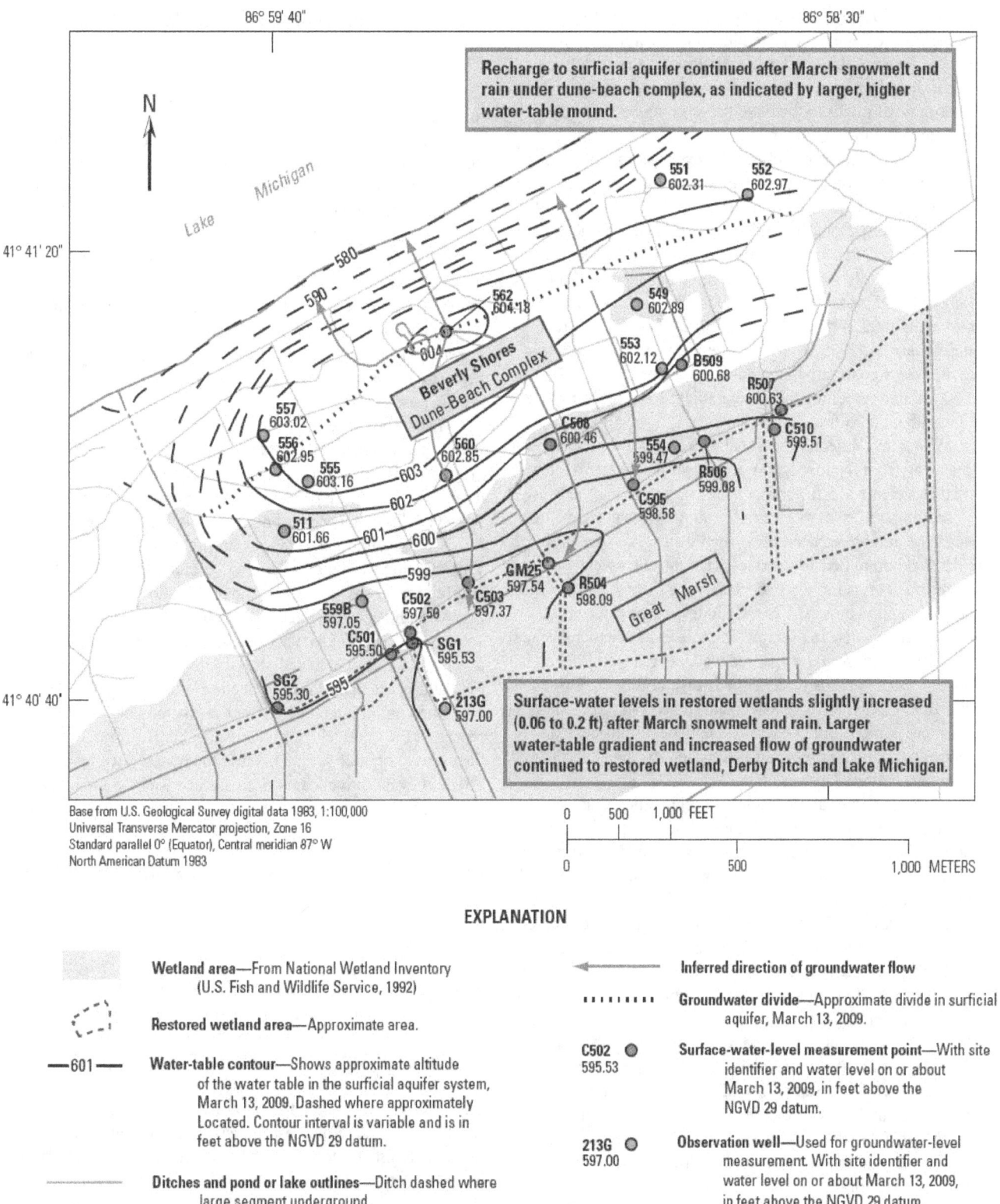

EXPLANATION

Wetland area—From National Wetland Inventory
(U.S. Fish and Wildlife Service, 1992)

Restored wetland area—Approximate area.

—601— Water-table contour—Shows approximate altitude
of the water table in the surficial aquifer system,
March 13, 2009. Dashed where approximately
Located. Contour interval is variable and is in
feet above the NGVD 29 datum.

Ditches and pond or lake outlines—Ditch dashed where
large segment underground.

Inferred direction of groundwater flow

⋯⋯⋯ Groundwater divide—Approximate divide in surficial
aquifer, March 13, 2009.

C502 ⚫ Surface-water-level measurement point—With site
595.53   identifier and water level on or about
March 13, 2009, in feet above the
NGVD 29 datum.

213G ⚫ Observation well—Used for groundwater-level
597.00   measurement. With site identifier and
water level on or about March 13, 2009,
in feet above the NGVD 29 datum.

Figure 41.   Water-table contours and directions of groundwater flow in the surficial aquifer, March 13, 2009, after 3.8 inches of precipitation during March 7–11, 2009, at the Indiana Dunes National Lakeshore station, northwestern Indiana. Contours indicate a rise in the water table beneath the dune-beach complex area between the Great Marsh and Lake Michigan and westward expansion of water-table mound toward Derby Ditch relative to conditions before the precipitation event.

The rapid growth of the water-table mound in the central part of the dune-beach complex after precipitation also increased water-table gradients and groundwater-flow rates toward discharge areas at the hydrologic boundaries (fig. 39). Water-table altitudes in the dune-beach complex typically increased within 1 to 6 hours after the start of rainfall. The rise in water-table altitude broadened the extent of the flat part of the water-table mound toward the hydrologic boundaries. This broadening increased the water-table gradient and groundwater-flow velocity from the mound toward the discharges at the marsh, Derby Ditch, and Lake Michigan. For example, the water-level gradients (the rate of change in water-level altitude along a flow direction) between wells 553 and 554 increased from about 0.0022 ft/ft on February 4, 2009, to 0.0032 ft/ft on February 11, 2009, after recharge from snowmelt and rain reached the water table in the dune-beach complex (fig. 39). Groundwater velocities for these gradients ranged from about 0.18 to about 0.26 ft/d.

Water levels decline seasonally because of groundwater flow away from the mound toward the surface-water bodies, and they decline at a faster rate after rainfall during the growing season because of evapotranspiration. More groundwater leaves the dune-beach complex when water levels are higher under the mound because surface-water levels typically do not rise as much or as rapidly except after the highest precipitation and snowmelt events. The corresponding surface-water-level increase in response to precipitation relates to whether surface-water discharge from the restored wetland and Derby Ditch is constrained at culverts under Broadway and Beverly Drive and from Derby Ditch to Lake Michigan. Increasing the capacity of present culverts under roadways may help limit development of temporarily higher surface-water levels in the restored wetland and along Derby Ditch.

Recharge from about 3.8 in. of precipitation that fell during March 7–11, 2009, also caused the water table to rise beneath the dune-beach complex with an additional westward expansion of the broad, relatively flatter part of the water-table mound toward Derby Ditch, as indicated by the 603-ft water-level contour (fig. 41). The cumulative effect of infiltration of snowmelt and precipitation from February 6 through March 13, 2009, produced a water-table shape and directions of flow (fig. 41) that were similar to those on September 15, 2008 (fig. 33). The March 13, 2009, water-table configuration represents the addition of about 7.2 in. of precipitation over a 5-week period from February 6 to March 13, 2009, as compared with the 3-day, 10.77-in., September 12–15, 2008, rainfall. Infiltration of snowmelt and precipitation from February 6 through March 13, 2009, also increased the size of the water-table mound, the gradient between wells in the mound and wells on the outer slope of the mound.

Effect of Evapotranspiration on Groundwater Levels

The effect of evapotranspiration on groundwater levels in the study area is shown in the general pattern of decreasing water-table altitudes from May to August of 2009 (figs. 24–25). Water-level altitudes in the dune-beach complex for wells 549, 551, 553, and 555 (fig. 17A) were lower from May to September 2009 despite receiving occasional rainfall during May through August (figs. 23, 26–27). Evapotranspiration in the region near Lake Michigan commonly is greater than the total amount of rainfall in the summer months (Beaty, 1994). Water levels in these wells reached their lowest values in September (wells 549 and 553) and October (well 551). Groundwater levels adjacent to the restored marsh in well 554 did not begin to decrease until mid-June 2009, as did surface-water levels (figs. 27–30).

The importance of evapotranspiration losses to daily declines in water-table altitudes is indicated by declines in groundwater levels during daylight hours but not during night hours (fig. 42). The daytime-only water-level decline was observed during a period of no recorded rainfall from both the Great Marsh and INDU stations during August 10–15, 2009. Transpiration from plants "pumps" groundwater out of the aquifer during daylight hours and appears to decline to near zero at night, causing a general decline of the water table over various days. Groundwater levels next to the restored wetland at well 554 also declined slightly more during daylight hours than at night but continued to decline at night (fig. 42). The continued nightly decline in water levels in well 554 may represent the relatively continuous discharge of groundwater through the organic sediments into the restored wetland or the effect of evaporative losses from surface water warmed by daytime sunlight and air temperatures in the restored wetland.

Effect of Water-Supply Changes on Groundwater Levels

The equivalent rise in groundwater altitude resulting from a hypothetical uniform annual application of water from the 2005-9 change to Lake Michigan water source was estimated to range from 0.4 to 0.5 ft. The estimate was prepared by assuming the uniform maximum application of 1.6 in/yr to 1.9 in/yr of lake-sourced water (table 8) to the approximately 400-acre area under the dune-beach complex in Beverly Shores, divided by the effective porosity of the surficial aquifer (0.3). This maximum estimate is based on several assumptions, including uniform water application over the entire year and over the entire 400 ac estimated area of water service in Beverly Shores, an assumed aquifer porosity, and no net consumptive loss of water by domestic use. In addition, greater water use during the drier summer season by spring and

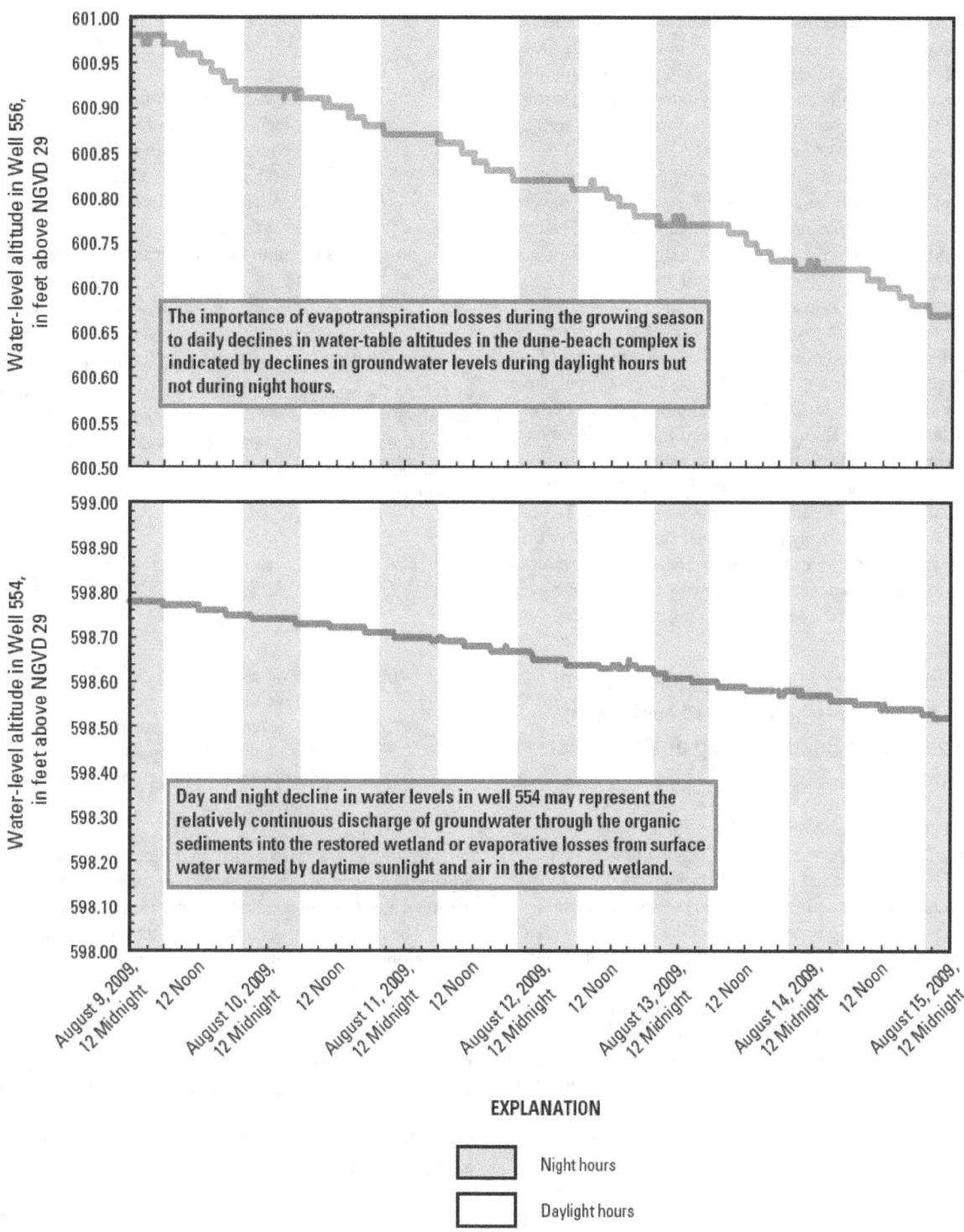

The importance of evapotranspiration losses during the growing season to daily declines in water-table altitudes in the dune-beach complex is indicated by declines in groundwater levels during daylight hours but not during night hours.

Day and night decline in water levels in well 554 may represent the relatively continuous discharge of groundwater through the organic sediments into the restored wetland or evaporative losses from surface water warmed by daytime sunlight and air in the restored wetland.

EXPLANATION

Night hours

Daylight hours

Figure 42.   Differences in the daytime decline and nightly stability (diurnal variation) of groundwater levels affected by evapotranspiration in well 551 in the dune-beach complex and the gradual decline of groundwater levels in well 554 adjacent to a restored part of the Great Marsh, northwestern Indiana, August 10–16, 2009.

summer seasonal residents and less water used during the wetter fall and winter seasons should result in smaller water-level increases attributable to water supply changes. The estimate provides a general indication of how groundwater levels might increase from infiltration of lake-sourced water to the aquifer through residential wastewater effluent.

Time-variable and concentrated application of septic-system effluent could cause small transient mounds to develop near septic systems during periods of higher sustained water use. Olyphant and Harper (1995, p. 13 and 15) describe temporary increases in the water table of about 2 in. at two observation wells, within a day after addition of about 53 ft^3 of water to a nearby dry well on June 9, 1994. This relatively instantaneous addition of water is higher than the range of average daily water use in Beverly Shores in 2009 from about 9.1 ft^3/d in December to 29.7 ft^3/d in August. Average daily water use in Beverly Shores was derived from data in table 6. The temporary increases in the water table of about 2 in. reported by Olyphant and Harper (1995) were less than the water-table rise they observed in response to a 3-day, 2–3 in. rainfall in July 1994. Daily and seasonal changes in flow from septic systems could focus the efflux of lake-sourced water to the aquifer when residences are seasonally occupied and as water use varies from day to day and throughout the day.

Simulated Effects of Recharge and Wetland Water Levels on Groundwater Levels in an Idealized Cross Section Using an Analytical Solution

An analytical solution describing a simplified cross section of groundwater flow (fig. 43; David Pollock, U.S. Geological Survey, Reston, Virginia, written commun., May 2009) was used to examine hydrologic processes affecting the water-table profile beneath the dune-beach complex. An analytical solution is a basic mathematical model that simulates aquifer response to water input and output. In this study, the hydrologic processes compared by the simulations were recharge from precipitation, recharge from Lake Michigan-derived domestic water use, and increased surface-water levels in the restored wetland. The simplified cross section (fig. 43) represents the unconfined surficial aquifer underlain by a flat relatively impermeable layer (confining unit) and bounded on each side by surface-water bodies—Lake Michigan on the north and the Great Marsh on the south. For these simulations, the restored marsh boundary is defined as approximately south of Beverly Drive (figs. 4, 17A and 17B). The simplified cross section is most similar to the hydrology of the east transect along the north-south section B–B' (figs. 4 and 15). In each simulation, the hydraulic conductivity of the aquifer and the recharge rate are spatially constant. The groundwater-flow equation used for the cross section is the same as presented by Bear (1979, equation 5-210, p. 180) with slightly different variable names:

$$\frac{d}{dx}\left((h-z_b)K\frac{dh}{dx}\right)+R=0$$

where,

x	is the distance, in feet, from the lake boundary;
R	is the recharge rate, in feet per day;
z_b	is the bottom altitude of the hypothetical aquifer, in feet above the datum;
h	is the water level altitude at point x, in feet above the datum;
$h-z_b$	is the saturated thickness of the aquifer, in feet; and
K	is the hydraulic conductivity of the aquifer, in feet per day.

The boundary conditions for the cross section are defined by setting the water-table altitude equal to the lake altitude on the left ($x = 0$) and the altitude of the marsh on the right ($x = L$). The analytical solution of the groundwater-flow equation subject to these boundary conditions (David Pollock, U.S. Geological Survey, Reston, Virginia, written commun., May 2009) is

$$h=\left(-\left(\frac{R}{K}\right)x^2+\left(\frac{RL}{K}\right)x+\left(\frac{(H_M-z_b)^2-(H_L-z_b)^2}{L}\right)x+(H_L-z_b)^2\right)^{\frac{1}{2}}+z_b$$

where,

HM	is the water-level altitude in the marsh, in feet above the datum;
HL	is the water-level altitude in the lake, in feet above the datum; and
L	is the distance, in feet from the lake to the marsh.

Aquifer characteristics and hydrologic conditions in the adjacent lake and marsh that were representative of local, assumed steady-state conditions were used in model simulations. The distance between the marsh and the lake used for the simulations (L) was 3,000 ft. The bottom altitude of the hypothetical aquifer (z_b) was assumed to be 575 ft above NGVD 29. The aquifer hydraulic conductivity (K) in all calculations was assumed to be 25 ft/d; this value was within the local range for aquifer sand (from 8 to 43 ft/d; Watson and others, 2002). The recharge rates from precipitation that were used for these simulations ranged from 11 in/yr (median recharge conditions; Beaty, 1994) to 22 in/yr (2009 conditions; table 6). The assumed initial surface-water level used to simulate pre-restoration conditions in the marsh for the simulations (596.8 ft) was similar to the median groundwater levels in the pre-restoration marsh at well GM25 during the typically high water-level conditions of October–April of 1979–89 (596.82 ft, table 9). The water level of the marsh used to simulate post-restoration conditions (597.9 ft) was similar to the median water level at site R504 for measurements made from 2007–9 (597.94 ft, table 9). The lake water level used for all simulations, 580 ft, was within the range of daily mean Lake Michigan water levels from June 1985 to September 1992 (578.58 to 583.10 ft; Greeman, 1995) and is similar to the maximum of the range of mean monthly

Idealized Section Across Aquifer

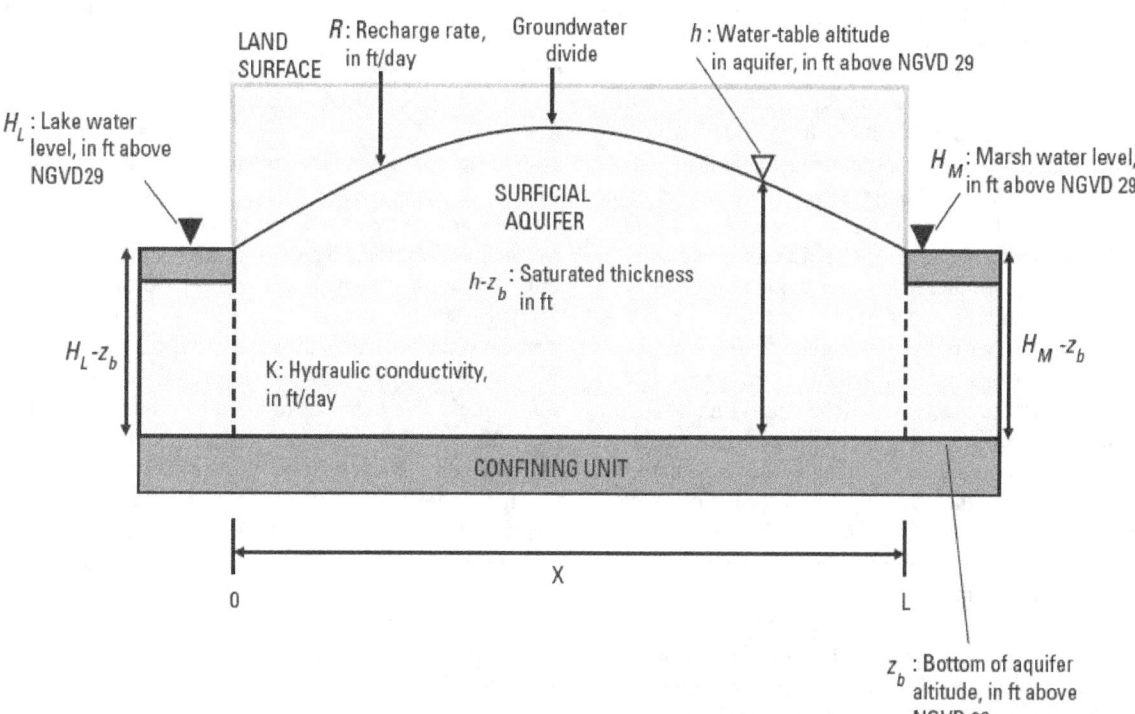

Figure 43. Diagram showing major hydrologic features of the idealized section across the surficial aquifer, underlying the area near section *B–B′*, that were simulated by using an analytical solution.

water levels during 2007-09 (575.35 ft in December 2007 to 579.91 ft in June 2009; National Oceanic and Atmospheric Administration, 2010).

Simulation Results and Relative Effects of Hydrologic Processes on Groundwater Levels

The analytical solution was able to represent a water-table crest for 2009 that was within the range before and after a major recharge event in early spring of 2009. The capacity of the analytical solution to represent local conditions was evaluated by comparing results of simulated recharge and water-level conditions typical of 2009 with the altitude of the water-table mound under the dune beach complex during late winter-early spring of 2009. The simulated water-table mound altitude was about 603.3 ft (fig. 44), during assumed conditions representing those in 2009 of 22 in/yr of recharge from precipitation (table 6), 1.9 in/yr of added recharge from Lake Michigan water supply (table 8), and a water-level altitude in a nearby part of the restored wetland of 597.9 ft (altitude at site R504, table 9). By comparison, the altitude of the crest of the water-table mound varied from about 602.8 ft on February 8, 2009, before a snowmelt to about 603.5 ft on March 6, 2009, and 604.2 ft on March 13, 2009 (figs. 38, 40–41).

The analytical solution underpredicted the altitude of the water-table mound during a period of more typical recharge in 1986 (11 in/yr) before wetland restoration. The water-table altitude at well 309 in a land-surface depression near the midpoint between the marsh and Lake Michigan under the dune-beach complex was from 600.57 ft on January 24, 1986, and 600.75 ft on May 15, 1986 (shown generally in fig. 17). Simulation results (fig. 44) indicate that recharge of about 15.2 in/yr would be needed to achieve an altitude of about 599.8 ft at the crest of the water-table mound. A relatively high precipitation amount from the prior November (about 7.3 in.) and daily high air temperatures above 32°F during January 11–12 and 16–19, 1986, immediately before the January 23, 1986, water-level measurement may have produced short-term rises in that water level. In addition, an irregular response to recharge distribution in land-surface depressions with smaller depths to the water table in the dune-beach complex or near other recharge sources (septic-system discharges) could also have produced a locally higher water-table altitude near well 309 than would be predicted by the simplified model simulation. These results, however, are sufficiently similar to the pattern of water-level increase across the dune-beach complex to indicate that the simplified model can be used in this setting to evaluate the relative importance of hydrologic processes affecting the water-table altitudes.

Model results indicate that increased recharge from precipitation and snowmelt was the principal cause of raised water levels in the dune-beach complex from 2006 to 2009 (fig. 44). Addition of an estimated 22 in/yr of recharge from precipitation in 2009 (11 in/yr above the median value) to the simulations resulted in about a 4-ft rise in groundwater level at the center of the water-table mound. Smaller amounts of additional recharge produced progressively smaller rises in the water-table altitude. Because the water-table altitude is held constant by the simulation at the restored wetland and lake boundaries, simulated water levels near those boundaries are not sensitive to changes in recharge. The simulated groundwater divide moves farther from the restored wetland and closer to Lake Michigan along the idealized cross section as the amounts of recharge are increased in the model (fig. 44). This result of the simulations reflects the increased contribution of recharge to the computed height of the water table in the analytical solution and the relatively diminished importance of the water level at the restored wetland boundary. The groundwater divides in the actual aquifer system are typically closer to Lake Michigan than model-simulated divides. Reasons may include the relative isolation of the surficial aquifer from surface-water levels in the wetland where overlain by organic sediments and because of possible focusing of recharge in low areas within the dune-beach complex that are closer to the lake.

Addition of 1.9 in/yr of recharge from the simulated change to a lake water source raised the groundwater divide altitude by only about 0.7 ft as compared with rises from recharge increases from precipitation in 2006, 2008, and 2009 (fig. 44), which were about 4 ft. Recharge from the addition of water from a lake water source was simulated as a uniform addition across the aquifer and increased water levels across most of the simulated cross section (fig. 45). The distribution of recharge from domestic water supply is one of the sources of uncertainty when comparing the model with the actual aquifer conditions. Based on these results, reductions in recharge from domestic water supply to the aquifer by routing wastewater outside the basin through sanitary sewers could produce relatively small decreases in groundwater levels.

The simulated effect of raised surface-water levels in the restored marsh on raised groundwater levels in the dune-beach complex was progressively smaller with distance from the marsh (fig. 45). For example, a simulated increase of 1.1 ft in water level in the restored marsh declined in effect to about a 0.75-ft increase in groundwater level within about 900 ft from the marsh and to about a 0.55 ft increase in groundwater level at about 1,500 ft from the marsh. By comparison, the simulated groundwater level rise at 900 ft from the marsh—about 0.7 ft—attributed to 1.9 in/yr of recharge from the change to a Lake Michigan water supply was comparable to the simulated rise resulting from the wetland restoration (fig. 45). Simulated rises in water-table altitude attributed to recharge

from precipitation at 900 ft from the restored wetland ranged from about 0.9 ft for the 2007 increase in precipitation-related recharge (2.5 in/yr) to about 3.8 ft for the 2009 increase in precipitation-related recharge (11 in/yr; fig. 45).

The effect of raised wetland water levels in the simulation was observed principally as progressively smaller increases in groundwater level with greater distance from the wetland. The size of groundwater-level changes in the actual dune-beach complex arising from wetland water-level changes would be so small as to be indistinguishable among the recharge-related causes of water-level increases and groundwater flooding. The analytical solution results indicate that much larger rises in wetland water level than those seen and analyzed during this study would be needed to produce groundwater-level rises that were comparable to those produced by precipitation increases since 2006.

Various characteristics of the surficial aquifer and dune-beach complex indicate that the actual groundwater-level change from raised wetland water levels is likely to be smaller than simulated changes. The organic sediments that separate the surficial aquifer in places from the wetland (figs. 14–15) appear to delay groundwater-level increases, as observed in groundwater-level changes from wells 559B and 554 relative to changes in surface-water levels (figs. 23, 27–29). This same characteristic may also delay groundwater discharge to the marsh. Ditches that drain groundwater from the dune-beach complex, such as measured at B509 and C508 (figs. 31, site locations on fig 17B), vary less than groundwater levels in nearby wells 549 and 553 (figs. 26–27), as would be expected. Water levels measured at B509 and C508 at the ditch appear to reflect the local water-table altitude. That ditch would moderate water-table changes so that the changes in surface-water level in the restored wetland represented in the simulation would not be propagated past the ditch. In that case, water-level changes at wells 549 and 553 would solely represent changes related to recharge from precipitation and from the change to a lake-water source and discharge by subsurface groundwater flow out of the study area, groundwater discharge to the ditch, and evapotranspiration.

Limitations of the Analytical Solution

The analytical solution provides a general, idealized description of how long-term (years) changes in hydrologic conditions can affect the water-table altitude of a hydrologic unit similar to that of the surficial aquifer. The simulations do not provide precise estimates of the exact rise and fall of the water table because of differences between the idealized and actual aquifer characteristics and conditions. The assumptions of the analytical solution, as applied to the idealized cross section, included the following similarities to and differences from the surficial aquifer.

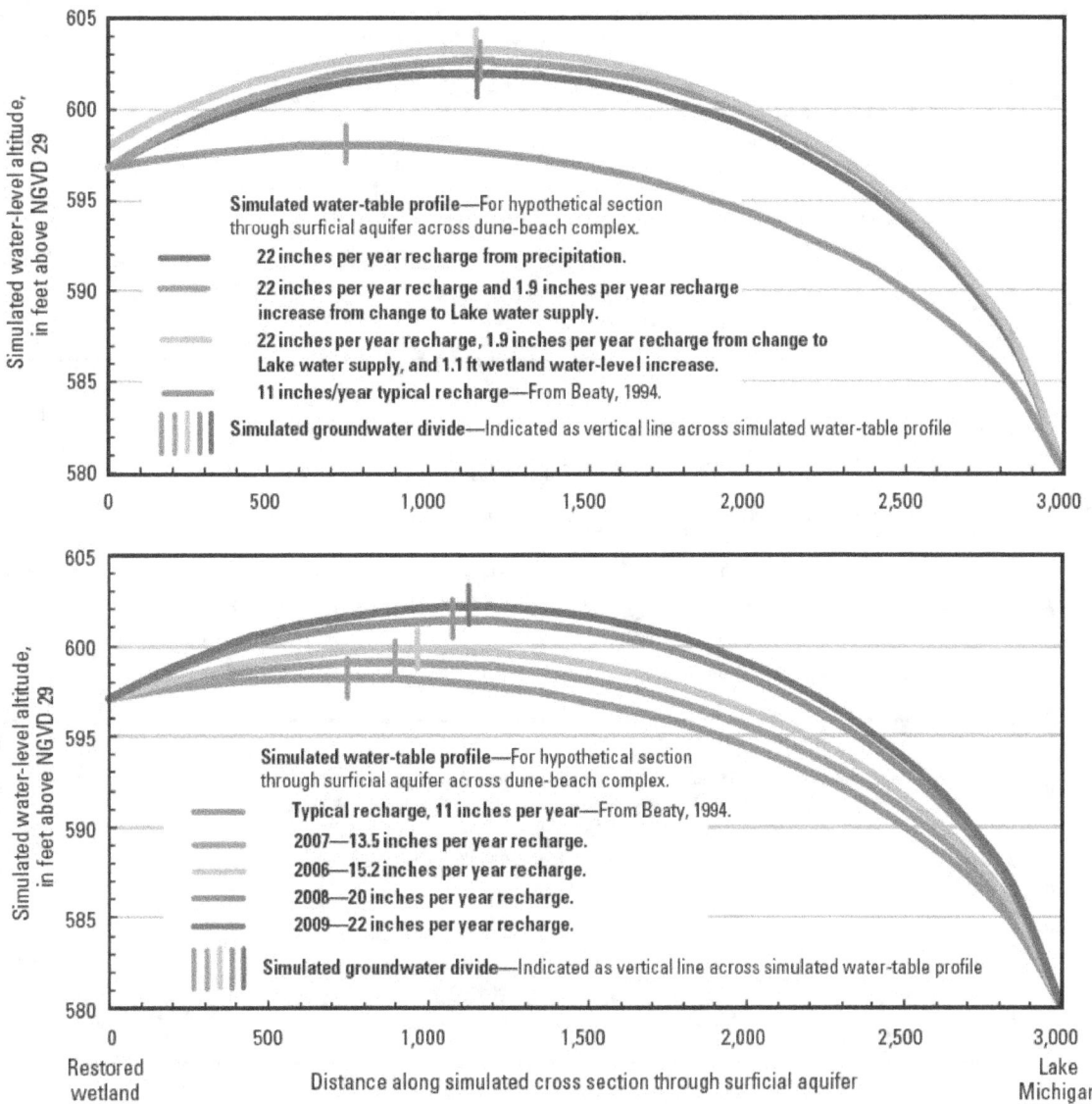

The simulated increase in recharge from precipitation in 2006-09 causes most of the groundwater level increase along the cross section except within about 200 ft of the restored wetland.

At the groundwater divide of the simulations, the simulated increase in recharge from 2009 precipitation increased simulated groundwater levels by about 4 ft.

Simulated recharge increase from the change to Lake Michigan water supply increased groundwater levels at the divide by about 0.7 ft. and the wetland water-level change increased simulated groundwater levels by about 0.5 ft.

The results represent an idealized setting. Groundwater level changes from precipitation would locally increase relative to these simulations because of recharge in shallow water-table areas in the dune-beach complex. Groundwater level changes from wetland water-level increases would be smaller than the simulated values because of organic sediments between the wetland and surficial aquifer.

Figure 44.   Simulated water-table altitudes for the simplified aquifer cross-section, based on varying recharge amounts from precipitation and cumulative effects from recharge, the change to a Lake Michigan water supply, and an increase in water level in the restored wetland, northwestern Indiana.

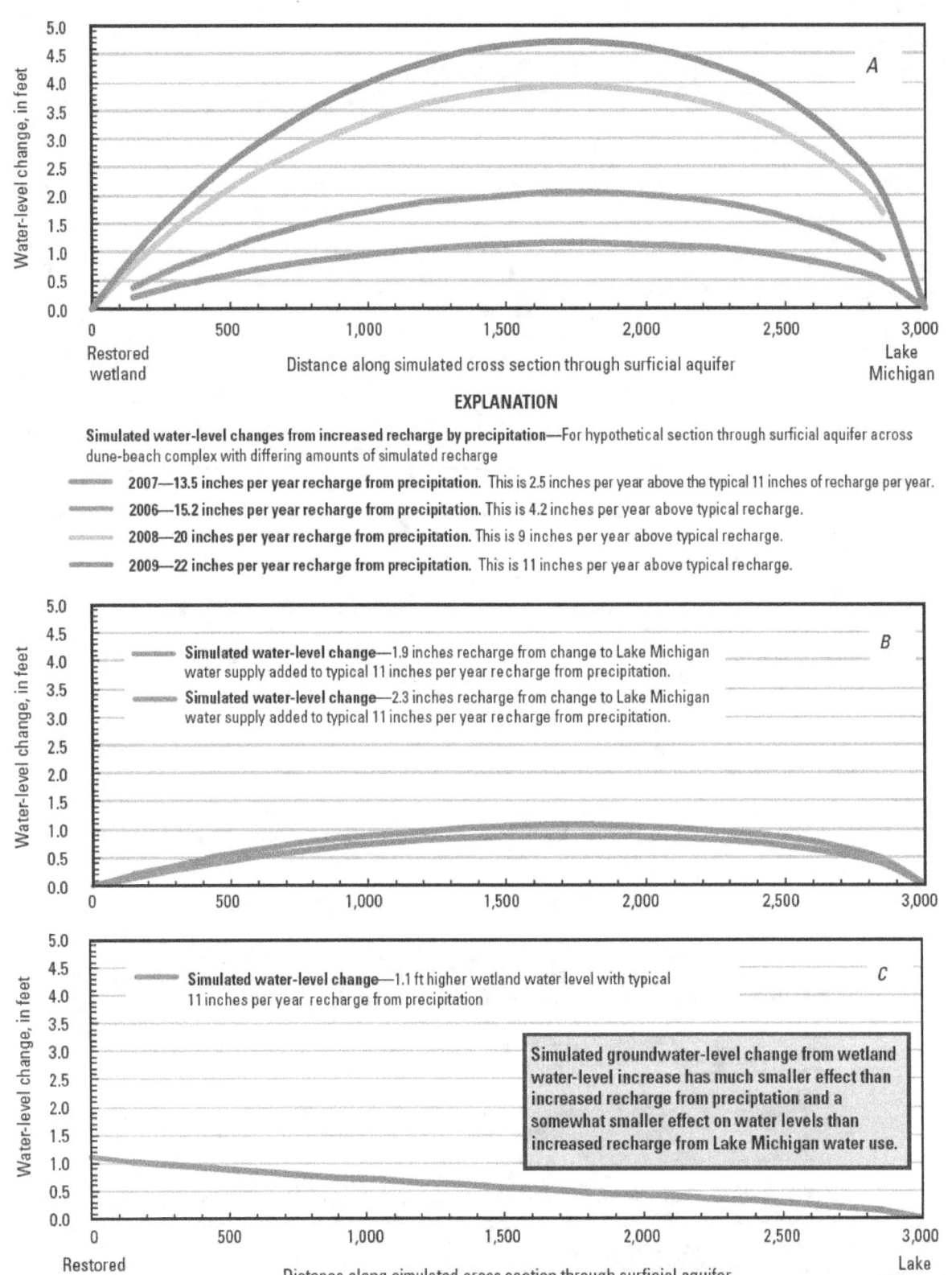

EXPLANATION

Simulated water-level changes from increased recharge by precipitation—For hypothetical section through surficial aquifer across dune-beach complex with differing amounts of simulated recharge

2007—13.5 inches per year recharge from precipitation. This is 2.5 inches per year above the typical 11 inches of recharge per year.

2006—15.2 inches per year recharge from precipitation. This is 4.2 inches per year above typical recharge.

2008—20 inches per year recharge from precipitation. This is 9 inches per year above typical recharge.

2009—22 inches per year recharge from precipitation. This is 11 inches per year above typical recharge.

Simulated water-level change—1.9 inches recharge from change to Lake Michigan water supply added to typical 11 inches per year recharge from precipitation.

Simulated water-level change—2.3 inches recharge from change to Lake Michigan water supply added to typical 11 inches per year recharge from precipitation.

Simulated water-level change—1.1 ft higher wetland water level with typical 11 inches per year recharge from precipitation

Simulated groundwater-level change from wetland water-level increase has much smaller effect than increased recharge from preciptation and a somewhat smaller effect on water levels than increased recharge from Lake Michigan water use.

Figure 45. Increases in water-table altitudes along the simplified aquifer cross-section for simulations based on the following assumptions: *A*, Increased amounts of precipitation-related recharge over 11 inches per year. *B*, Increased recharge amounts from the local use of water supplied from Lake Michigan. *C*, An increase in water level in the restored wetland relative to pre-restoration conditions.

The analytical solution, despite its numerical precision, represents a simplified version of the surficial aquifer under the dune-beach complex between the restored wetland and Lake Michigan that is not entirely accurate. For example, the following items describe simplifications of the surficial aquifer and hydrologic system that are inherent in the analytical solution.

1. The analytical solution represents the surficial aquifer as at steady state or a stable condition in response to changes in recharge and discharge from groundwater. In reality, the aquifer is in a continuous transient state of change in response to changes in recharge and discharge. The following are examples of transient changes that affect groundwater and surface-water levels in this study area:

    a. *Large rainfall of an inch or more or snowmelt equivalent to an inch or more of precipitation.*

    b. *Daily and seasonal variation of use of Lake Michigan water supply. That water use is seasonally larger during spring, summer, and autumn months.*

    c. *Evapotranspiration during spring, summer, and early autumn. Results from this study indicate that relatively larger amounts of precipitation-derived recharge during periods of low evapotranspiration have more effect on groundwater levels in the surficial aquifer in the dune-beach complex than do precipitation increases during periods of high evapotranspiration.*

2. Hydraulic properties of the simulated aquifer are assumed to be uniform (homogenous) and isotropic. At the scale of this application, this condition generally applies for the surficial aquifer. The lower permeability organic sediments near the marsh, however, are a heterogeneity that is not represented in the analytical solution. Communication of water-level changes in the marsh with the surficial aquifer depends on the extent to which groundwater can flow through the organic sediments. To the extent that the organic sediments are less permeable than the aquifer sediments, these sediments also would limit flow from the marsh to the aquifer.

3. The hypothetical aquifer in the analytical solution is assumed to have a uniform depth and infinite extent perpendicular to the hypothetical cross section. The "infinite extent" condition applies to the extent that groundwater flows parallel to the section through the aquifer toward the hydrologic boundaries (lake and marsh). The analytical solution describes a condition that is similar to the area near section *B–B′*, except where the ditch near site B509 conveys groundwater discharge out of the section. Similarly, the analytical solution describes conditions that do not apply to areas near section *A–A′*, where groundwater flows across and out of the cross section southwestward toward a discharge along Derby Ditch.

4. The aquifer in the idealized cross section is underlain by impermeable material. This assumed condition is generally met because the surficial aquifer is considerably more permeable than the underlying confining unit.

5. For these calculations, the altitude of the base of the aquifer is assumed to be constant–575 ft. The actual surficial aquifer has a sloping bottom and a variable thickness, which can cause calculated water levels to deviate from actual values.

6. The simulated aquifer is assumed to be bounded on each end by two parallel surface-water boundaries with surface-water levels—in this case, Lake Michigan and the Great Marsh—that are connected to water levels in the aquifer. Deviations from this condition, such as where the wetland and the surficial aquifer are separated by less conductive organic sediments, can cause surface-water-level changes in the wetland and groundwater-level changes in the surficial aquifer to lag each other in time.

7. The simulated aquifer is assumed to receive recharge that is uniformly applied and that uniformly infiltrates to the water table. Recharge to areas in topographic depressions with smaller depths to the water table, such as near well 562, would produce increased water-table altitudes after precipitation that were higher than those simulated by the model. In addition, recharge from effluent derived from lake-supplied water is 'simulated' as areally homogenous, but in fact has a point distribution. Actual water-table altitudes would be lowered relative to the calculated values by drainage that redirects groundwater away from the section. Examples of this drainage are the ditch near site B509 and the tile drain near site R570.

Despite the limitations described above, the analytical solution permitted comparison of the relative effect of major changes to recharge and pre-and post-restoration wetland water levels on groundwater levels in an idealized representation of the surficial aquifer beneath the Beverly Shores area. A more precise evaluation of the groundwater-flow system would require a distributed, three-dimensional digital model to represent the variation of aquifer properties, thickness, hydraulic conductivity and time-dependent changes in recharge from precipitation, snowmelt and water use and groundwater and surface-water levels in response to changing recharge and discharge conditions.

Summary and Conclusions

The potential for high groundwater levels to cause wet basements (groundwater flooding) is of concern to residents of communities in northwestern Indiana. Changes in recharge owing to precipitation increases during 2006–9, water-level changes from restoration of nearby wetlands in the Great Marsh in 1998–2002, and changes in recharge due to the end of most groundwater withdrawals since 2005 at Beverly Shores, Ind., have been suspected as possible factors in development of increased groundwater levels in the surficial aquifer. From 2007 through 2010, the U.S. Geological Survey and the National Park Service investigated the relative effect of natural and human-affected hydrologic processes on changes in groundwater and surface-water levels and groundwater-flow directions in an unconfined surficial aquifer beneath nearby parts of a dune-beach complex.

The study was done in two parts of the surficial aquifer in the dune-beach complex, in a 1.7-mi^2 area at Beverly Shores, Ind., near a restored wetland in the Great Marsh and in a 0.1-mi^2 undeveloped part of the Indiana Dunes National Lakeshore at Howe's Prairie. Both are within 0.5 mi of Lake Michigan in northwestern Indiana. The surficial aquifer is an unconfined aquifer composed of dune, beach, and lacustrine sands, with a saturated thickness ranging from about 8 to 20 ft in the areas under study. A confining unit of till, glaciolacustrine silty clay, and clay deposits underlies the surficial aquifer. The surficial aquifer is part of a dune-beach complex of topographically high dunes and intervening intradunal wetlands. Groundwater recharges the aquifer and flows from a groundwater divide under the dune-beach complex to discharges on the north at Lake Michigan, to the south near restored wetland cells in the Great Marsh, and to the west into a part of Derby Ditch that drains the Great Marsh and flows to Lake Michigan. Less conductive organic sediments somewhat separate the surficial aquifer from surface water in the wetland.

The years 2006–9, following restoration of wetland hydrology in the Great Marsh (1998-2002), were the wettest 4-year period for precipitation in a combined 1952–2009 record of data from two stations in and near the Indiana Dunes National Lakeshore. Annual precipitation totals at the Indiana Dunes station in 2006–9 were 50.11, 44.89, 55.75, and 43.88 in/yr, respectively, and were substantially greater than the median precipitation of 36.35 in/yr. Estimates of recharge to groundwater from precipitation in these years were 15.2 in/yr in 2006, 13.5 in/yr in 2007, 22 in/yr in 2008, and 20.5 in/yr in 2009. Recharge rates to groundwater in 2006–9 were higher than the typical 11 in/yr because of the higher amounts of annual precipitation received as large events and during non-growing-season months in 2008–9.

The net recharge to groundwater from domestic water use has increased since the 2005 change in Beverly Shores to a Lake Michigan-derived water source but is smaller than the increase derived from precipitation changes. The cessation of groundwater withdrawal from the surficial aquifer for domestic supply and replacement of that with a source from Lake Michigan, coupled with a continued efflux of domestic wastewater to septic systems, causes a net increase in recharge to the surficial aquifer. The volume of Lake Michigan water supplied to Beverly Shores ranged from 2,300,900 ft^3/yr in 2006 to 2,719,700 ft^3/yr in 2009 and equated to an estimated increase in recharge from 1.6 to 1.9 in/yr.

Groundwater levels were measured in 2 existing and 13 new observation wells and surface-water levels were measured at 14 sites at least weekly during 2008–9 and occasionally during 2007–8 for the study area at Beverly Shores. Groundwater levels measured from 1985–2004 and in 2009 at five wells in an interdunal wetland at Howe's Prairie also were used for this study to identify groundwater-level responses in the surficial aquifer from precipitation changes.

Wetland water levels were higher and varied less after wetland restoration as compared with before. Surface-water levels in the restored wetland near the midpoint of the study area were up to about 1.1 ft higher in 2007–9 after wetland restoration as compared with seasonally wet periods from October to April in 1979–89. Groundwater levels next to an adjacent downstream wetland cell near Broadway during winter and spring wet periods before wetland restoration were similar to those after wetland restoration. The data indicate that similar surface-water levels and ponding of water were probable during winter and spring wet periods before and after restoration, though the periods of ponded water were likely longer during 2007–9.

Groundwater levels in 2008–9 at well 552, near the middle of the dune-beach complex, were similar to those in historical data from nearby prior wells during 1982–89 before water-supply and wetland-restoration changes. Water levels in prior well 309 through the 1981–88 period of record were consistently above the base of a hypothetical 6-ft-deep basement used to represent basements in the residential area of Beverly Shores. These very limited data from wells in one small part of the dune-beach complex indicated that the potential for basement flooding from high groundwater levels in nearby low areas in the dune-beach complex predated wetland restoration and water-supply changes in 2005.

High groundwater-level altitudes at Howe's Prairie in 2009 near a natural wetland were caused by infiltration of the high precipitation amounts received by the area from December 2008 through April 2009. Groundwater levels from Howe's Prairie in 2009 were similar to the highest water levels measured in the late summer and autumn months of 1990; these water levels followed a 7.77-in. rainfall on August 18, 1990, and other relatively high rainfalls. Similarly, high water levels in 2009 from the Howe's Prairie wells were also measured during the spring months in 1991 and 1993. Annual precipitation amounts in 1990 were the highest on record, and in 1993 were the fourth highest on record. The Howe's Prairie groundwater-level data indicate that recharge from similarly high precipitation amounts in 2008–9 was also a likely cause of high groundwater levels in other parts of the dune-beach complex, such as at Beverly Shores.

Groundwater-level fluctuations lasting days to weeks in the dune-beach complex were superimposed on a seasonal high water-table altitude; that seasonal high water-table altitude began with the recharge from snowmelt and rain in February 2009 and continued through July 2009. Minimum groundwater levels were observed near the end of the growing season and before appreciable precipitation in or close to non-growing-season months, such as before a 10.77-in. September 12–15, 2008, precipitation event and in mid-September 2009, before precipitation events later that month. Most of the relatively larger rises in water level of 0.5 ft or more were in wells in the dune-beach complex during or after rainfall l snowmelt events. Groundwater-level fluctuations in these wells varied over a relatively narrow range of about 2 to 3 ft, with no well having net fluctuations greater than 4 ft for the 2008–9 period of record. Smaller, shorter-term rises in water level after individual rain events persisted over hours to less than 1 week.

Nearly immediate, sharp rises in groundwater levels followed by relatively slow recessions were observed after 1–2 day precipitation events of 1 in. or more in September, October, and December 2008 and in early March and late October 2009. Sharp rises in water level followed by more rapid recessions were observed during the growing season of April through September 2009.

Water levels in several wells that are in the dune-beach complex (511, 549, 551, 553, 556, and 562) and adjacent to the Great Marsh (559B, 554) were within 0 to 6 ft of the land surface. Water levels at these sites indicate that basements in nearby residences could be within a depth that requires dewatering to maintain dry conditions. All but one of these wells (556) were in or near an interdunal pond (562), the restored wetland (554), or interdunal wetlands in relatively low areas of the dune-beach complex (511, 549, 551, 553 and 559B).

Surface-water-level fluctuations during this study generally varied over a narrower range than groundwater fluctuations, approximately from 1 to 1.5 ft. Surface-water levels increased more than this range after the 10.77-in. rainfall in mid-September 2008; water levels increased from about 2.5 ft to as much as 4.5 ft the most at downstream sites and increased about 0.5 ft or less at upstream sites in the dune-beach complex.

Perennial mounding of the water table and groundwater-flow directions in the surficial aquifer under the dune-beach complex indicate that the recharge that created the mound originates within the complex and not from flow from the adjacent hydrologic boundaries: the restored wetland, Lake Michigan, and Derby Ditch. Relatively greater increases in groundwater levels beneath the center of the dune-beach complex after precipitation and snowmelt as compared to smaller increases near the surface water at the complex's periphery indicate that recharge from infiltrating precipitation and wastewater cause the mounding. Infiltrating precipitation causes most seasonal and episodic rises in groundwater levels beneath the dune-beach complex. For example, after about 10.77 in. of rain fell during September 12–15, 2008, water levels rose in all wells but increased the most beneath the dune-beach complex. Relatively larger water-level rises were in wells at relatively lower areas inside the dune-beach complex, where the depth to the water table was lowest: next to an interdunal pond at well 562 (about 2.3 ft) and in an interdunal wetland area at well 549 (about 2.1 ft). These increases in water-table-mound altitude under the dune-beach complex recurred in 2008–9 in response to the largest rain events of 2.75 in. or more and to a snowmelt. Surface-water drainage through Derby Ditch to Lake Michigan may have limited some changes in surface-water levels and some groundwater levels along the western boundary of the study area at Derby Ditch.

Rapid water-level rises in the restored wetland after precipitation do not likely have an effect on groundwater flooding elsewhere in the dune-beach complex. Time delayed and smaller groundwater-level rises in wells 554 and 559B indicate a delaying effect on groundwater-level changes in and near the restored wetland from less conductive organic deposits in the subsurface near the marsh. Groundwater levels adjacent to the marsh at the two sites did not rise rapidly in response to short-term—less than 1 week—changes to surface-water levels in the restored part of the marsh.

Groundwater is removed from the surficial aquifer through discharge to adjacent surface water, to intradunal drains, and by evapotranspiration. Groundwater-level gradients from the water table mound to wells next to discharges increase after rainfall and snowmelt events and recede slowly as groundwater discharges from the aquifer. Evapotranspiration is responsible for part of the general pattern of decreasing water-table altitudes observed in continuous and weekly measurements from May to August 2009.

Results of a simplified, steady-state cross-sectional model of groundwater flow also indicate that increased recharge from precipitation and snowmelt was the principal cause of raised water levels in the dune-beach complex from 2006 to 2009. Addition of a simulated 1.9 in/yr of recharge from Lake Michigan-derived water source raised the altitude of the groundwater divide by about 0.7 ft. Rises in that divide caused by increased recharge from precipitation in 2006, 2008, and 2009 were about from 2 to 4 ft. A 1.1 ft simulated increase from a post-restoration marsh water-level change decreased in effect with distance to about a 0.75-ft increase at about 900 ft from the marsh and to about 0.55 ft within about 1,500 ft of the marsh (The simulated wetland margin was south of Beverly Drive—a major east-west street. Well 553 is about 900 ft from the restored wetland.) The simulated groundwater-level increase from a wetland water-level increase of 1.1 ft was much smaller across the idealized cross section than the increased groundwater level from precipitation-related recharge in 2006–9. The simulated groundwater-level increase from a wetland water-level increase of 1.1 ft was also somewhat smaller than the increased groundwater level derived from increased recharge owing to the change by Beverly Shores to a Lake Michigan water source.

The simulation results, however, represent an idealized setting; actual water-level changes would differ. Groundwater

Water-Level Fluctuations and Flow Directions near Beverly Shores, Indiana: Questions and Answers

- *How do water levels compare from periods before wetland restoration (1978–89) and afterward (2007–9) in the Great Marsh near Beverly Shores?*

  They appear to compare closely. Water levels were high during 2007–9, but they also were high in the past in a nearby area during times of above-normal precipitation and before marsh restoration.

  (See section "Water-Level Fluctuations Before and After Wetland Restoration at Beverly Shores" in the main report.)

- *How did groundwater levels change in a similar nearby but unrestored wetland area (Howe's Prairie) during the same period?*

  Groundwater-level changes in the surficial aquifer at the unrestored Howe's Prairie area followed nearly identical patterns as those at Beverly Shores, implying that precipitation and associated recharge were the predominant influences on groundwater levels.

  (See section "Groundwater-Level Fluctuations Without Wetland Restoration at Howe's Prairie" in the main report.)

- *Do groundwater levels in the dune-beach complex indicate a potential for groundwater flooding of basements?*

  Yes, but water levels near land surface seem to be particularly likely in low-lying places. Among study-area wells where water was within 6 ft of land surface, all but one was either in an interdunal pond, within the restored wetland, or in interdunal low areas in the dune-beach complex.

  (See section "Groundwater-Level Fluctuations Near Beverly Shores After Wetland Restoration, 2007–9" in the main report.)

- *What are the directions of groundwater flow relative to the Great Marsh and Lake Michigan, how do they change before and after precipitation, and what do the flow directions indicate about processes that affect groundwater levels?*

  An elongated mound in the water table, with an orientation that parallels the lake shore, controls directions of groundwater flow in the surficial aquifer. The crest of the mound in the water table is in the beach-dune complex between Lake Michigan and the Great Marsh. Flow directions are consistently from the mound outward to Lake Michigan, the Great Marsh, or drainage structures near the perimeter of the complex, both before and after major precipitation events. This pattern indicates the predominance of precipitation and recharge as controls of groundwater levels and the relatively insignificant effect on groundwater levels of the small surface-water-level rise resulting from wetland restoration.

  (See section "Groundwater-Flow Directions in the Surficial Aquifer" in the main report.)

- *Which natural and human-affected hydrologic processes most affect groundwater and surface-water levels and their short-term (daily-monthly) and longer term (seasonal) fluctuation in the dune-beach complex?*

  - *Recharge from precipitation (additive).*—Increased recharge from precipitation and snowmelt was the principal cause of raised water levels in the dune-beach complex from 2006 to 2009. A simulated addition of 22 in/yr of recharge from precipitation in 2009—11 in/yr more than the median value—resulted in about a 4-ft rise at the crest of the water-table mound. Smaller amounts of recharge produced smaller rises in water-table altitude.

(See sections "Groundwater-Flow Directions in the Surficial Aquifer" and "Simulation Results and Relative Effects of Hydrologic Processes on Groundwater Levels" in the main report.)

- *Decreased groundwater withdrawal from the change in water supply (additive).*—The rise in groundwater altitude resulting from a hypothetical uniform annual application of water from the 2005 change to Lake Michigan supply was small, about half a foot. A different simulation computed a 0.7-ft groundwater-level rise at about 900 ft from the marsh from the change to Lake Michigan water supply. That result was similar to the rise resulting from the simulated rise from wetland restoration.
(See sections "Groundwater-Flow Directions in the Surficial Aquifer" and "Simulation Results and Relative Effects of Hydrologic Processes on Groundwater Levels" in the main report.)

- *Recharge of the surficial aquifer by seepage from the restored wetland (additive).*—The effect of raised surface-water levels in the restored marsh on raised groundwater levels in the dune-beach complex decreased with distance from the marsh in an idealized simulation. A simulated increase of 1.1 ft in water level in the restored marsh declined in effect to about a 0.75-ft increase in groundwater level within about 900 ft from the marsh and to about a 0.55-ft increase in groundwater level at about 1,500 ft from the marsh. Organic sediments that separate the surficial aquifer in places from the wetland appear to delay groundwater-level increases relative to changes in surface-water levels.
(See sections "Groundwater-Flow Directions in the Surficial Aquifer" and "Simulation Results and Relative Effects of Hydrologic Processes on Groundwater Levels" in the main report.)

- *Groundwater discharge from the surficial aquifer to the restored wetland (removal).*— Groundwater levels decline seasonally because of groundwater flow away from the mound toward adjacent surface-water bodies: the restored wetland, Derby Ditch and Lake Michigan.
(See section "Groundwater-Flow Directions in the Surficial Aquifer" in the main report.)

- *Evapotranspiration (removal).*—Groundwater levels decline at a faster rate after rainfall during the growing season because of evapotranspiration. The importance of evapotranspiration losses to daily declines in water-table altitudes is indicated by declines in groundwater levels during daylight hours but not during night hours.
(See section "Effect of Evapotranspiration on Groundwater Levels" in the main report.)

- *Other drainage, such as tile drains (removal).*—A ditch in the dune-beach complex moderates water-table changes so that changes in surface-water level in the restored wetland would not be propagated past the ditch. The effect of tile-drain flow restoration on rates of groundwater-level decline in July and August 2009 could not be distinguished in the 2008–9 water-level data.
(See sections "Groundwater-Flow Directions in the Surficial Aquifer" and "Wetland and Drainage Effects on Water-Level Fluctuations" in the main report.)

level increases from precipitation-related recharge may locally be larger relative to these simulations where the recharge reaches the aquifer more effectively due to a thinner unsaturated zone. Groundwater-level changes from wetland water-level increases would be smaller than the simulated values because of the delay in head transmitted across the lower hydraulic conductivity organic sediments between the base of the restored marsh and the surficial aquifer.

These results indicate that increased groundwater levels and groundwater flooding reported for basements in the dune-beach complex in 2006–9 principally related to increased recharge from precipitation. Changes in groundwater levels that may be produced by increased supply of Lake Michigan water or by increased water levels in adjacent restored wetlands since 2002 appear to be much smaller and cannot be distinguished from those caused by increased precipitation. Results from the simulations indicate that reductions in recharge from domestic water supply to the aquifer by routing wastewater outside the basin through sanitary sewers could produce relatively small decreases in groundwater levels.

References Cited

Alley, W.M., Reilly, T.E., and Franke, O.L., 1999, Sustainability of Ground-Water Resources: U.S. Geological Survey Circular 1186, 86 p.

Bayless, E.R., and Arihood, L.D., 1996, Hydrogeology and simulated ground-water flow through the unconsolidated aquifers of northeastern St. Joseph County, Indiana: U.S. Geological Survey Water-Resources Investigations Report 95–4225, 47 p.

Bear, Jacob, 1979, Hydraulics of Groundwater: New York, McGraw-Hill, 567 p.

Beaty, J.E., 1994, Water resource availability in the Lake Michigan Region, Indiana: Indiana Department of Natural Resources, Division of Water, Water Resource Assessment 94-4, 257 p.

Buszka, P.M., Fitzpatrick, John, Watson, L.R., and Kay, R.T., 2007, Evaluation of ground-water and boron sources by use of boron stable-isotope ratios, tritium, and selected water-chemistry constituents near Beverly Shores, northwestern Indiana, 2004: U.S. Geological Survey Scientific Investigations Report 2007–5166, 46 p.

Changnon, S.J., 1968, Precipitation climatology of Lake Michigan Basin: Illinois State Water Survey Bulletin 52, 46 p., accessed May 10, 2010, at *http://www.isws.illinois.edu/pubdoc/B/ISWSB-52.pdf.*

Cole, K.L., and Taylor, R.S., 1995, Past and current trends of change in a dune prairie/oak savanna reconstructed through a multiple-scale history: Journal of Vegetation Science, v. 6, p. 399-410.

Duwelius, R.F., Yeskis, D.J., Wilson, J.T., and Robinson, B.A., 2002, Geohydrology, water quality, and simulation of ground-water flow in the vicinity of a former waste-oil refinery near Westville, Indiana, 1997–2000: U.S. Geological Survey Water-Resources Investigations Report 01–4221, 161 p.

Freeze, R.A., and Cherry, J.A., 1979, Groundwater: Englewood Cliffs, N.J., Prentice-Hall, 604 p.

Grannemann, N.J., Hunt, R.J., Nicholas, J.R. Reilly, T.E., and Winter, T.C., 2000, The importance of ground water in the Great Lakes region: U.S. Geological Survey Water-Resources Investigation Report 00-4008, 19 p.

Greeman, T.K., 1995, Water levels in the Calumet aquifer and their relation to surface-water levels in northern Lake County, Indiana, 1985–92: U.S. Geological Survey Water-Resources Investigations Report 94–4110, 61 p.

Hartke, E.J., Hill, J.R., and Reshkin, Mark, 1975, Environmental geology of Lake and Porter Counties, Indiana—An aid to planning: Bloomington, Indiana Geological Survey Special Report 11, 57 p.

Isiorho, S.A., Beeching, F.M., Stewart, P.M., and Whitman, R.L., 1994, Seepage measurements from Long Lake, Indiana Dunes National Lakeshore: Environmental Geology, V. 28, no. 2, p. 99–105.

Midwestern Regional Climate Center, 2010, Products and services—Online data—MACS: accessed May 11, 2010, at *http://mcc.sws.uiuc.edu/prod_serv/prodserv.htm.*

National Oceanic and Atmospheric Administration, 2010, Tides and Currents, Calumet Harbor, IL 9087044: accessed May 16, 2010, at *http://tidesandcurrents.noaa.gov/geo.shtml?location=9087044.*

National Park Service, 2009, Great Marsh Restoration at Indiana Dunes National Lakeshore: accessed May 25, 2010, at *http://www.nps.gov/indu/planyourvisit/upload/great_marsh_restoration_pdf.pdf.*

National Weather Service, 2009, A wet start to 2009: Chicago, Ill., Weather Forecast Office, accessed May 12, 2010, at *http://www.crh.noaa.gov/news/display_cmsstory.php?wfo=lot&storyid=27347&source=2.*

National Weather Service, 2008, Tropical summary message—Public advisory number 56 for remnants of Ike: Camp Springs, Md., NWS Hydrometeorological Prediction Center, accessed May 12, 2010, at *http://www.hpc.ncep.noaa.gov/tropical/2008/IKE/IKE_56.html.*

Neff, B.P., Piggott, A.R., and Sheets, R.A., 2005, Estimation of shallow ground-water recharge in the Great Lakes Basin: U.S. Geological Survey Scientific Investigations Report 2005–5284, 20 p.

Olyphant, G.A., and Harper, Denver, 1995, Chemistry and movement of septic-tank absorption-field effluent in the dunes area, Lake and Porter Counties, Indiana: Indiana Geological Survey Open-File Report 95–8, 83 p.

Scheeringa, Ken, 2002, Climate of Indiana: Indiana State Climate Office, accessed April 28, 2011 at *http://iclimate.org/narrative.asp*

Shaffer, K.H., and Runkle, D.L., 2007, Consumptive water-use coefficients for the Great Lakes Basin and climatically similar areas: U.S. Geological Survey Scientific Investigations Report 2007–5197, 191 p.

Shedlock, R.J., Cohen, D.A., Imbrigiotta, T.E., and Thompson, T.A., 1994, Hydrogeology and hydrochemistry of dunes and wetlands along the southern shore of Lake Michigan, Indiana: U.S. Geological Survey Open-File Report 92–139, 85 p.

Shedlock, R.J., Wilcox, D.A., Thompson, T.A., and Cohen, D.A., 1993, Interactions between ground water and wetlands, southern shore of Lake Michigan, USA: Journal of Hydrology, v. 141, p. 127–155.

Shedlock, R.J., and Harkness, W.E., 1984, Shallow ground-water flow and drainage characteristics of Brown Ditch basin near the East Unit, Indiana Dunes National Lakeshore, Indiana: U.S. Geological Survey Water-Resources Investigations Report 83–4271, 37 p.

Thompson, T.A., 1987, Sedimentology, internal architecture and depositional history of the Indiana Dunes National Lakeshore and State Park: Bloomington, Ind., Indiana University, unpublished Ph.D. thesis, 129 p.

Thompson T.A., 1992, Beach-ridge development and lake-level variation in southern Lake Michigan: Sedimentary Geology, vol. 80, p. 305–318.

Thornthwaite, C.W., 1948, An approach toward a rational classification of climate: Geographical Review, v. 38, p. 55–94.

U.S. Census Bureau, 2010, Beverly Shores town, Indiana, American FactFinder: accessed April 28, 2010, at *http://factfinder.census.gov*.

U.S. Fish and Wildlife Service, 1992, National Wetland Inventory National Wetland Inventory Polygons by County in Indiana: 1:2M, Polygon Shapefile, accessed March 9, 2011, at *http://inmap.indiana.edu/dload_page/hydrology.html*

Watson, L.R., Bayless, E.R., Buszka, P.M., and Wilson, J.T., 2002, Effects of highway-deicer application on ground-water quality in a part of the Calumet aquifer, northwestern Indiana: U.S. Geological Survey Water-Resources Investigations Report 01–4260, 148 p.

Westenbroek, S.M., Kelson, V.A., Dripps, W.R., Hunt, R.J., and Bradbury, K.R., 2010, SWB—A modified Thornthwaite-Mather Soil-Water-Balance code for estimating groundwater recharge: U.S. Geological Survey Techniques and Methods 6–A31, 59 p.

Winter, T.C., 1999, Relation of streams, lakes, and wetlands to groundwater flow systems: Hydrogeology Journal, v. 7, p. 28–45.